PUBLISHER COMMENTARY

We print NASA's handbooks and standards for the convenience of those that use them on a daily basis. We print all of these a full 8 ½ by 11 with large text so they are easy to read. Yes, color books are expensive to print so unless the information relies on the use of color for proper interpretation or understanding, we print most books in black and white to keep the cost down. All these documents are available for download for free from NASA, however printing them all over a network printer would take days.

Why buy a book you can download free? We print this so you don't have to.

All these books are available for free download from the government web site. Some are available only in electronic media. Some online docs are missing pages or barely legible.

We at 4th Watch Publishing are former government employees, so we know how government employees actually use the standards. When a new standard is released, an engineer prints it out, punches holes and puts it in a 3-ring binder. While this is not a big deal for a 5 or 10-page document, many NIST documents are over 100 pages and printing a large document is a time-consuming effort. So, an engineer that's paid $75 an hour is spending hours simply printing out the tools needed to do the job. That's time that could be better spent doing engineering. We publish these documents so engineers can focus on what they were hired to do – engineering. It's much more cost-effective to just order the latest version from Amazon.com

If there is a standard you would like published, let us know. Our web site is www.usgovpub.com

www.usgovpub.com

List of Other NASA Publications Available on Amazon.com:

NASA-STD-5001B	Structural Design and Test Factors of Safety for Spaceflight Hardware
NASA-STD-5006A	General Welding Requirements for Aerospace Materials
NASA-STD-5008B	Protective Coating of Carbon Steel, Stainless Steel, and Aluminum on Launch Structures, Facilities, and Ground Support Equipment
NASA-STD-5009A	Nondestructive Evaluation Requirements for Fracture-Critical Metallic Components
NASA-STD-5012B	Strength and Life Assessment Requirements for Liquid-Fueled Space Propulsion System Engines
NASA-STD-5019A	Fracture Control Requirements for Spaceflight Hardware
NASA-STD-5005D	Standard for The Design and Fabrication of Ground Support Equipment
NASA-HDBK-8739.21	Workmanship Manual for Electrostatic Discharge Control
NASA-HDBK 8739.23A	NASA Complex Electronics Handbook for Assurance Professionals (Color)
NASA-HDBK-8719.14	Handbook for Limiting Orbital Debris (Color)
NASA-HDBK-8709.22	Safety and Mission Assurance Acronyms, Abbreviations, and Definitions
NASA-HDBK-7009	NASA Handbook for Models and Simulations: An Implementation Guide For NASA-STD-7009 (Color)
NASA-HDBK-8739.19-2	Measuring and Test Equipment Specifications NASA Measurement Quality Assurance Handbook – Annex 2
NASA-HDBK-8739.19-3	Measurement Uncertainty Analysis Principles and Methods NASA Measurement Quality Assurance Handbook – Annex 3
NASA-HDBK-8739.19-4	Estimation and Evaluation of Measurement Decision Risk NASA Measurement Quality Assurance Handbook – Annex 4
NASA RCM	Reliability-Centered Maintenance Guide for Facilities and Collateral Equipment

www.usgovpub.com

NASA HANDBOOK

Estimation and Evaluation of Measurement Decision Risk

NASA Measurement Quality Assurance Handbook – ANNEX 4

Measurement System Identification: **Metric**

July 2010

National Aeronautics and Space Administration
Washington DC 20546

This page intentionally left blank.

DOCUMENT HISTORY LOG

Status	Document Revision	Approval Date	Description
Baseline		2009-07-13	Initial Release *(JV/L4)*
	Revalidated	2018-03-01	Baseline revalidated.

This document is subject to reviews per Office of Management and Budget Circular A-119, Federal Participation in the Development and Use of Voluntary Standards (02/10/1998) and NPD 8070.6, Technical Standards (Paragraph 1.k).

This page intentionally left blank.

FOREWORD

This publication provides principles and methods for the analysis and management of the uncertainty that exists in measurements. The principles and methods described herein support Agency objectives for assuring measurement accuracy and are applicable for use in all NASA functions for which measurements and decisions based on measurements are involved.

This Annex to NASA-HDBK 8739.19 is approved for use by NASA Headquarters and NASA Centers, including Component Facilities. This document may be referenced on contracts as a guidance or training publication.

Comments and questions concerning the contents of this publication should be referred to the National Aeronautics and Space Administration, Director, Safety and Assurance Requirements Division, Office of Safety and Mission Assurance, Washington, DC 20546.

Requests for information, corrections, or additions to this NASA HDBK shall be submitted via "Feedback" in the NASA Technical Standards System at http://standards.nasa.gov or to National Aeronautics and Space Administration, Director, Safety and Assurance Requirements Division, Office of Safety and Mission Assurance, Washington, DC 20546.

Bryan O'Connor   13 JULY 2010
Chief, Safety and Mission Assurance   Approval Date

This page intentionally left blank.

TABLE OF CONTENTS

List of Figures

List of Tables

Preface

Measurement data are used to make decisions that impact all areas of technology. Whether measurements support research, design, production, or maintenance, ensuring that the data supports the decision is crucial. The quality of measurement data affects the consequences that follow measurement-based decisions. Negative consequences from measurement results can range from wasted resources to loss of mission or life. Historically, to ensure measurement data supported decisions, selection and calibration of MTE was the emphasis for measurement quality assurance. With ever increasing technology requirements, the emphasis needs shift to understanding and controlling measurement decision risk in all areas of technology.

AS9100C defines risk as, "*An undesirable situation or circumstance that has both a likelihood of occurring and a potentially negative consequence.*" The focus of measurement quality assurance is to quantify, and/or manage the "*likelihood*" of incorrect measurement-based decisions. When doing so, there must be a balance between the level of effort and the risks resulting from making an incorrect decision. In balancing the effort versus the risks, the decision (direct risk) and the *consequences* (indirect risk) of the measurement must be considered.

1. **Direct Risk:** This risk is directly associated with the measurement data and impacts the decisions involving a measurement (e.g., accept, reject, rework, scrap).

2. **Indirect Risk:** This risk affects the quality or performance of end products which stem from measurements. In other words, this is the "*consequence*" of an incorrect decision. This type of risk may not be evident until after the product is in service.

This Handbook provides tools for estimating and evaluating the measurement decision risk. Measurement decision risk analysis can be used to mitigate consequences associated with noncompliance to specifications and/or requirements which are validated through measurement. The principles and methods recommended in this Handbook may be used to design and support a quality measurement program. From this foundation, good measurement data can support better decisions.

Acknowledgements

The principal author of this publication is Dr. Howard T. Castrup of Integrated Sciences Group (ISG), Inc. of Bakersfield, CA under NASA Contract.

Additional contributions and critical review were provided by:

Suzanne G. Castrup
Integrated Sciences Group

Greg Cenker
Southern California Edison/EIX

Scott M. Mimbs
NASA/Kennedy Space Center

James Wachter
MEI Co., NASA Kennedy Space Center

Dr. Dennis H. Jackson
U.S. Naval Warfare Assessment Station

Jerry L. Hayes
Hayes Technology

Miguel A. Decos
Rohmann Services, Inc., NASA Johnson Space Center, TX

Del Caldwell
Caldwell Consulting

Perry King
Bionetics, Inc., NASA Kennedy Space Center, FL

Mihaela Fulop
SGT, NASA Glenn Research Center, OH

This Publication would not exist without the recognition of a priority need by the NASA Metrology and Calibration Working Group and funding support by NASA/HQ/OSMA and NASA/Kennedy Space Center.

Purpose and Scope

This document is Annex 4 of NASA-HDBK-8730.19 [40]. It provides guidelines for computing metrics by which conformance testing for test and calibration processes can be evaluated. Chief among these metrics are measurement decision risks. These include the risk of incorrectly accepting a nonconforming equipment attribute and the risk of incorrectly rejecting a conforming attribute. The former is called False Accept Risk and the latter is termed False Reject Risk. A principal objective of this Annex is to accumulate and focus the body of knowledge of measurement decision risk to provide uniform guidance in the application of this knowledge to estimate and manage such risk, particularly as it relates to NASA tests and calibrations.

In addition to providing guidelines for estimating and managing measurement decision risks, this Annex addresses the evaluation of these risks in terms of their impact on both operating costs and on the cost of the undesirable consequences of incorrect test decisions. Further, the guidelines provided herein can be applied to equipment and system design issues involving decisions regarding accuracy requirements and available support capability.

The principles and methods presented in this Annex are applicable to any conformance test that relies on measured values to provide information on which to base acceptance or rejection decisions.

Executive Summary

This document is one of a series of Annexes to NASA-HDBK-8730.19 [40], applying the science of analytical metrology to the assurance of measurement quality.[1] Measurement quality assurance (MQA) extends to many areas of technology management, including ensuring the accuracy of equipment attributes in the manufacturing process, the accuracy of the results of experiments, and the control of measurement uncertainty in calibration and testing. MQA also addresses the need for making correct decisions based on measurement results and offers the means to limit the probability of incorrect decisions to acceptable levels. This probability is termed "measurement decision risk."

The Test and Calibration Hierarchy

MQA is exercised in the use and management of primary measurement standards, working standards, calibration systems, test systems and end items, as shown in Figure 1.

Acceptable measurement decision risk at any level of the hierarchy is governed by the need to ensure that end items emerging from testing have an acceptable probability of performing as expected.

Ensuring End Item Performance

In the mid '80s, the problem of relating end item performance to calibration and testing capability was addressed in a rigorous way in which calibration, testing, and other equipment life

---

[1] NCSLI RP-1 [39], NCSLI RP-5 [38], NCSLI RP-12 [26].

cycle costs, as well as the costs of degraded end item utility, were minimized by applying an integrated model, referred to as the ETS model.[2] The ETS model applies measurement science and cost modeling to establish functional links from level to level in the test and calibration hierarchy with the aim of ensuring acceptable end item performance in a cost-effective manner. The development of such links involves the application of measurement uncertainty analysis, measurement decision risk analysis, calibration interval analysis, life cycle cost analysis and end item utility modeling. It also involves the acquisition, interpretation, use, and development of equipment attribute specifications.

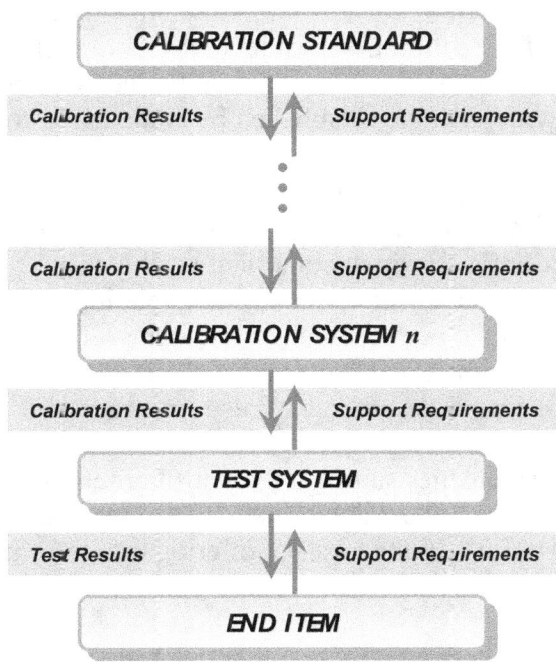

Figure 1. The Test and Calibration Support Hierarchy.

The support hierarchy links end item support to the capabilities at the hierarchy's various levels of measurement. The hierarchy also serves as a diagnostic to identify end item support requirements and to flag needed new support capabilities.

Since ETS was first introduced, these areas of endeavor have been updated. Updates are incorporated in a revised ETS model that provides the backbone for end-to-end MQA, as documented in Chapter 2 and Annex 1 of NASA-HDBK-8730.19 [40].

The use of such a model is critically important in the modern technological environment. The pressures of the competitive international marketplace and of national aerospace, energy, environmental and defense performance objectives and reliability requirements have led to a situation in which end item performance tolerances rival and, in some cases, exceed the best accuracies attainable at the primary standards level. In such cases, the interpretation of test and calibration data and the management of test and calibration systems require that the subtleties of the test/calibration process be accounted for and their impact on end item performance and support costs be quantified.

---

[2] Equipment Tolerancing System [13, 36, 27].

The Role of Measurement Decision Risk

The testing of a given end item attribute by a test system yields a reported in- or out-of-tolerance indication (referenced to the attribute's tolerance limits), an adjustment (referenced to the attribute's adjustment limits) and a "post-test" in-tolerance probability or "measurement reliability." Similarly, the results of calibration of the test system attribute are a reported in- or out-of-tolerance indication (referenced to the test system attribute's test limits), an attribute adjustment (referenced to the attribute's adjustment limits) and a beginning-of-period measurement reliability (referenced to the attribute's tolerance limits). The same sort of data results from the calibration of the calibration system and accompanies calibrations up through the hierarchy until a point is reached where the item calibrated is itself a calibration standard and, finally, a primary standard with its value established by comparison to the SI.

This hierarchical "compliance testing" is marked by measurement decision risk at each level. This risk takes two forms: **False Accept Risk**, in which non-compliant attributes are accepted as compliant, and **False Reject Risk**, in which compliant attributes are rejected as non-compliant. The effects of the former are possible negative outcomes relating to the accuracies of calibration systems and test systems and to the performance of end items. The effects of the latter are costs due to unnecessary adjustment, repair and re-test, as well as shortened calibration intervals and unnecessary out-of-tolerance reports or other administrative reaction.

The use of an integrated model enables setting risk control criteria that optimize end item MQA support. If an integrated model is not available, risk criteria may still be established based on the criticality of the test or calibration. In the absence of criticality information, nominal criteria may be used [33].

However the criteria are established, an effective methodology for estimating and evaluating measurement decision risk is required. The description of this methodology is presented in this document.

Chapter 1: Introduction

In recent years, ISO, NIST and ANSI/NCSL guidelines have been developed that provide a framework for analyzing and communicating measurement uncertainties. These guidelines constitute a major step in building a common analytical language for both domestic and international trade. This common language is important for ensuring that equipment tolerance limits and other measures of uncertainty have the same meaning for both buyer and seller.

The motivation for developing such a common understanding derives primarily from the need to control *false accept risk*. False accept risk is the probability that out-of-tolerance product or other attributes are perceived to be in-tolerance. False accept risk constitutes a measure of the quality of a measurement process as viewed by individuals external to a calibration or testing organization. The higher the false accept risk, the greater the chance of products not meeting performance or other quality expectations, resulting in returned goods, loss of reputation, litigation and other undesirable outcomes. In a commercial context, individuals external to a calibration or testing organization are labeled "consumers." For this reason, false accept risk has traditionally been called *consumer's risk*.

In the late 1970s it was realized that false accept risk could be viewed from two perspectives. On the one hand, there is the view of the calibrating or testing agency. In this view, false accept risk is considered to be the probability that out-of-tolerance attributes will be erroneously accepted by the testing or calibration process. The false accept risk defined by this view is not conditional on an acceptance or rejection event. Accordingly, it is referred to as "unconditional false accept risk" or *UFAR*.

The second viewpoint is that of the user of the tested or calibrated equipment. From this viewpoint, the important risk is the probability that accepted attributes will be out-of-tolerance. Since the false accept risk defined by this view is conditional on the acceptance of a tested or calibrated item, it is referred to as "conditional false accept risk" or *CFAR*.

A counterpart to false accept risk is false reject risk. False reject risk is the probability that in-tolerance attributes are perceived to be out-of-tolerance. False reject risk is a measure of the quality of a measuring process as viewed by individuals within the measuring organization; the higher the false reject risk, the greater the chances for unnecessary re-work and re-test. In a commercial context, a measuring organization is labeled the "producer," and false reject risk is called *producer's risk*. False reject risk has been given the acronym *FRR*.

False accept risk and false reject risk, taken together, are referred to as *measurement decision risk*.

1.1   Why Compute Risks?

1.1.1   Measurement Quality Metrics

The technical community has long been aware of the need to control measurement uncertainties to levels that are commensurate with the intended use of tested or calibrated equipment. This awareness has been accelerated recently with the introduction of national and international standards and guidelines.[3] Much of this has been motivated by the need to enforce standards of conformance for manufactured goods and to ensure acceptable performance of advanced military, space and private sector technologies.

The objectives of these concerns can be attained by linking the tolerances of tested or calibrated attributes to the uncertainty in the test or calibration measurement process. This link is forged through the analysis of measurement decision risk.

Given this, it can be asserted that false accept and false reject risks can be regarded as the principal metrics by which the quality of a test or calibration process can be evaluated. The upshot of this is that the uncertainties in measurements made by testing or calibration laboratories be regarded as operational attributes rather than as isolated entities. In other words, uncertainties need to be controlled to achieve desired levels of measurement decision risk.

1.1.2 Economics

In addition to providing quality metrics, risks constitute variables that relate to both the cost of testing and calibration and the cost of deploying or shipping nonconforming attributes. As stated earlier, the higher the false accept risk, the greater the chance for unsatisfactory product performance and associated undesirable outcomes.

As for false reject risk, there are two associated cost consequences. The first is simply that money spent correcting false rejects is money that is largely wasted. The second consequence is that, if the rejected item is an article of MTE, the out-of-tolerance condition is noted in the service history for the item, with the possible result that the item's calibration interval will be unnecessarily shortened. This may constitute a major source of wasted funding in that intervals are more sensitive to out-of-tolerances than to in-tolerances and, secondly, the more often an item is calibrated, the greater its support cost.

Linking decision risks to costs will be covered more extensively later.

1.2 Factors Affecting Risks

In testing or calibration, both false accept and false reject risks are affected by several factors. While the relationships between risks and these factors are complex, it can be easily appreciated, at least qualitatively, that the factors described in the following make up the list:

Reference attribute Accuracy

The reference attribute is the attribute of an item of MTE or a measurement system that makes the measurement associated with a given test or calibration step. Measurement attribute accuracy is stated in terms of either the tolerance limits of the reference attribute or confidence limits associated with the calibration of this attribute by a higher-level standard.

---

[3] See, for example, ISO/IEC 17025 [44] and ANSI/NCSL Z540.3 [33].

Reference attribute In-Tolerance Probability

This is the probability that the reference attribute is in-tolerance or within the confidence level for its stated confidence limits.

Measurement Process Uncertainty

This is the standard uncertainty in the total error of a testing or measurement process, including the reference attribute bias uncertainty.

Measurement Process Error Distribution

This is the specific mathematical form of the probability distribution of the total combined error of the testing or calibration process.

UUT attribute Tolerance Limits

The unit-under-test (UUT) attribute is the attribute under test or calibration. Its tolerance limits are the limits that it is tested or calibrated to. They may be thought of as the published tolerance limits for the attribute. In the NASA Handbook, an attribute's tolerance limits are sometimes called *performance limits*.

UUT attribute Test Limits

These are limits of acceptance or rejection of the UUT attribute during testing or calibration.

UUT attribute In-Tolerance Probability

This is the probability that the UUT attribute is in-tolerance prior to testing or calibration. As we will see later, this variable is a major driver of measurement decision risk.

UUT attribute Error Distribution

This is the specific mathematical form of the probability distribution of biases in the UUT attribute population.

1.3   Reader's Guide

While this document is intended to provide the information necessary for technical personnel to apply measurement decision risk analysis to measurement procedures, management and other non-technical personnel may benefit from some of the document's material. The following summarizes the content of each chapter and appendix to assist readers of all technical levels and interests.

Chapter 1.0 - Introduction

Chapter 1 introduces the topic of measurement decision risk, discusses several motivations for computing risks, and identifies factors that affect measurement decision risk.

Chapter 2.0 - Uncertainty Analysis

Chapter 2 provides an overview of uncertainty analysis methods and concepts. These methods and concepts are required to compute measurement uncertainties that affect measurement decision risk. Chapter 2 should be read as a prerequisite to Chapters 3 and 4.

Chapter 3.0 - Probability Relations

Chapter 3 covers the basics of applied probability and examines the mathematical relationships that comprise the foundation of measurement decision risk analysis.

Chapter 4.0 - Computing Risk

Chapter 4 presents methods for computing risks. Three principal alternatives are described; the "classical method" found in much of the measurement decision risk literature, the "Bayesian method" applicable to controlling measurement decision risk in response to the real-time results of testing or calibration, and the "confidence level method" that serves as a pseudo risk control tool to be used in the absence of information needed to estimate the bias uncertainty of the unit under test.

Chapter 4 utilizes the test and calibration measurement scenarios developed in Appendix A.

Chapter 5.0 - Compensating Measures

Chapter 5 covers methods that apply to compensating for risks in cases where risks are unacceptable for intended applications. Chapter 5 also discusses how what we observe during testing or calibration is affected by measurement decision risk. Based on this, approaches for managing periodic test and calibration are discussed. These approaches are expanded on in Appendix E.

Appendix A – Uncertainty Analysis

Appendix A provides supporting material for Chapter 2. It is intended as an abridged version of Annex 3, included in the present Annex for convenience. As such, it offers relatively brief discussions on the fundamentals of uncertainty analysis, descriptions of error sources often encountered in testing and calibration, relevant probability density functions for attribute biases and other error sources, a mathematical definition of uncertainty and a general method for combining uncertainties in measurement. Estimating uncertainty in multivariate measurements is also discussed as is the development of uncertainty estimates for four basic test and calibration scenarios.

Readers interested in a full development of uncertainty analysis concepts, principles and methods are encouraged to consult Handbook Annex 3 and NCSLI's RP-12 [26].

Appendix B – Test and Calibration Quality Metrics

In Appendix B, a set of probability functions are developed for evaluating the quality of testing and calibration. These metrics include the usual false accept and false reject risks, along with other metrics that may be of interest, such as the risk that non-conforming attributes will be accepted, the probability that conforming attributes will be accepted and the probability that non-conforming attributes will be rejected.

Appendix C – Introduction to Bayesian Measurement Decision Risk Analysis

Appendix C provides an introduction to this important risk control methodology at the level of high school algebra. The tools of single-variable calculus are encountered as brief excursions into the mathematics needed to develop a few necessary relations.

Appendix D – Derivation of Key Bayesian Expressions

Appendix D develops Bayesian methods of risk control and other applications for the more serious readers. The mathematics are complete and unvarnished and the methodology may require a reformation of one's conventional concepts of testing and calibration. Given the benefits of the Bayesian methodology, it is worth the effort.

Appendix E – True vs. Reported Probabilities

In testing and calibration, what we see is not always what we've got. Attributes that are perceived as being in-tolerance may be out-of-tolerance and those perceived as being out-of-tolerance may actually be in-tolerance. In compiling statistics on end of period (EOP) in-tolerance percentages, we suffer from the fact that our "reported" in-tolerance probabilities are not the same as our "true" in-tolerance probabilities. Appendix E deals with this problem and offers a means of correction for it. This has implications for both evaluating measurement processes and for setting periodic calibration intervals.

Appendix F – Useful Numerical Algorithms

Computing measurement decision risks and other measurement quality metrics requires the use of iterative algorithms embedded in computer programs. Two of the most useful of these are a bisection algorithm and a Gauss-quadrature integration algorithm. Both are documented in Appendix F.

Appendix G – Calibration Feedback Analysis

The problem of what course of action to take when an attribute of a test system is found out-of-tolerance during calibration is one that has plagued quality control organizations for decades. No generally accepted methodology exists for developing an effective response to such events. Accordingly, responses range from doing nothing to providing reports to test equipment users for every recorded out-of-tolerance. The former can be irresponsible, while the latter is probably overkill and often places responsibility on someone who may not be conversant with measurement quality control methodology.

Appendix G provides a feedback analysis methodology that evaluates the probability that the out-of-tolerance test system attribute may have falsely accepted a subset of the end items that it tested. The methodology applies the principles of uncertainty growth and the estimation of false accept risk to determine if some corrective action should be taken. Certain data, needed for the analysis are described. It turns out that these data are readily available and inexpensive to maintain.

Appendix H – Risk-Based End of Period Reliability Targets

EOP reliability targets need to be established to ensure that the in-tolerance probability of MTE attributes is held to an acceptable minimum during use. What this in-tolerance probability should be is a question that has historically not been satisfactorily answered. Appendix H aims at correcting this deficiency by offering an answer in terms of measurement decision risk control. The answer includes a solved-for reliability target and a method for determining reliability target confidence limits.

Appendix I – Set Theory Notation for Risk Analysis

As a consequence of the fact that risk analysis requires the application of probability theory, some of the notation in this document involves the use of the symbols of mathematical logic and set theory. While some readers find such notation to be merely elements of a convenient language of discourse, others may be perplexed by the seemingly arcane symbology. Appendix I, while not constituting a comprehensive essay on the topic will, hopefully, provide a mapping of notation to meaning that will be useful in reading and understanding this document.

Appendix J: Post-Test Risk Analysis

Appendix J explores the topic of post-test performance degradation of calibrated instruments. In calibration and testing, the measurement results are given either in a report of the measured value with details regarding the measurement uncertainty, or in terms of the results of a conformance test. Ordinarily, calibration results *do not* include uncertainties for the effects of transport and handling, effects of environmental conditions, or drift over time. In some cases, uncertainties arising from these effects may be greater than the reported uncertainty or tolerance of the test or calibration.

Appendix K: Derivation of the Degrees of Freedom Equation

Appendix K provides the derivation of the degrees of freedom equation. The amount of information used to estimate the uncertainty in a given error is called the *degrees of freedom*. The degrees of freedom is required to employ an uncertainty estimate in computing confidence limits commensurate with some desired confidence level.

1.4 Terms and Definitions

The terms and definitions employed in this document are designed to be understood across a broad technology base. Where possible, terms and definitions have been taken from internationally recognized standards and guidelines in the fields of testing and calibration, as well as the International vocabulary of Basic and General Terms in Metrology (VIM) [12].

Term	Definition
a posteriori value	A value determined after taking measurements.
a priori value	A value assumed before measurements are taken.
Accuracy	Closeness of agreement between a declared or measured value of a quantity and its true value.
	In terms of instruments and other measuring devices, accuracy is defined as the conformity of an indicated value to the true value or, alternatively, the value of an accepted standard.

Term	Definition
Accuracy Ratio	See Test Accuracy Ratio.
Analog Signal	A quantity or signal that is continuous in both amplitude and time.
AOP	The in-tolerance probability for an attribute averaged over its calibration or test interval. The AOP measurement reliability is often used to represent the in-tolerance probability of an attribute for a measuring item whose usage demand is random over its test or calibration interval.
Arithmetic Mean	The sum of a set of values divided by the number of values in the set.
Artifact	A physical object or substance with measurable attributes.
Asymmetric Distribution	A probability distribution in which deviations from the population mode value of one sign are more probable than deviations of the opposite sign. Asymmetric distributions have a nonzero coefficient of skewness.
Attribute	A measurable characteristic, feature, or aspect of a device, object or substance.
Attribute Bias	A systematic deviation of an attribute's nominal or indicated value from its true value.
Average	See Arithmetic Mean.
Average-Over-Period (AOP)	See AOP.
Beginning-of-Period (BOP)	The start of a calibration or test interval.
Between Sample Sigma	A standard deviation representing the variation of values obtained for different samples taken on a given quantity. See also Within Sample Sigma.
Bias	A systematic discrepancy between an indicated, assumed or declared value of a quantity and the quantity's true value. See also Attribute Bias and Operator Bias.
Bias Uncertainty	The uncertainty in the bias of an attribute or error source quantified as the standard deviation of the bias probability distribution.
BOP Reliability	The in-tolerance probability for an MTE attribute at the start of its calibration or test interval.
Calibration	An operation in which the value of a measurand is compared with a corresponding value of a measurement reference, resulting in (1) a physical adjustment of the measurand's value, (2) a documented correction of the measurand's value, or (3) a determination that the measurand's value is within its specified tolerance limits.

Term	Definition
Calibration Interval	The scheduled interval of time between successive calibrations of an equipment parameter or attribute.
Characteristic	A distinguishing trait, feature or quality.
Coefficient Equation	An equation that expresses the partial derivative of a parameter value equation with respect to a selected error source. This equation is used to compute a sensitivity coefficient for the selected error source.
Combined Error	An error comprised of a combination of two or more error sources.
Combined Uncertainty	The uncertainty in a combined error.
Component Error	See Error Component.
Component Uncertainty	The product of the sensitivity coefficient for an error component and the standard uncertainty of the error component or a constituent error source.
Computation Error	The error in a quantity obtained by computation. Normally due to machine round-off error, error in values obtained by iteration, or errors due to the use of regression models. Sometimes applied to errors in tabulated physical constants.
Computed Mean Value	The average value of a sample of measurements.
Conditional False Accept Risk (*CFAR*)	The probability that an equipment attribute, accepted by conformance testing, is out-or-tolerance.
Confidence Level	The probability that a set of limits will contain a given error.
Confidence Limits	Limits that bound errors of a given error source with a specified confidence level.
Conforming	Applied to an attribute if its true value lies within or on the range of values bounded by its tolerance limits.
Conformance Test	The measurement of an attribute value or bias in order to decide conformance or nonconformance with specifications.
Containment Limits	Limits that are specified to contain either an attribute value, an attribute bias, or other measurement process error.
Containment Probability	The probability that an attribute value or the error in the measurement of this value lies within specified containment limits.
Correlation	A probability relationship between two or more random variables.
Correlation Analysis	An analysis that determines the extent to which two random variables influence one another. Typically the analysis is based on ordered pairs of values. In the context of measurement uncertainty analysis, the random variables of interest are error sources or error components.

Term	Definition
Correlation Coefficient	A measure of the extent to which two errors are linearly related. A function of the covariance between the errors of two variables. Correlation coefficients range from minus one to plus one.
Covariance	The expected value of the product of the deviations of two random variables from their respective means. The covariance of two *independent* random variables is zero. See Independent Error Sources.
Coverage Factor	A factor used to express an error limit or expanded uncertainty as a multiple of the standard uncertainty.
Cross-correlation	The correlation between two error sources of two different error components in a multivariate analysis.
Cumulative Distribution Function	A mathematical function whose values $F(x)$ are the probabilities that a random variable assumes a value less than or equal to x. Synonymous with Distribution Function.
Degrees of Freedom	A statistical quantity that is related to the amount of information available about an uncertainty estimate. The degrees of freedom signifies how "good" the estimate is and serves as a useful statistic in determining appropriate coverage factors and computing confidence limits and other decision variables.
Deviation from Nominal	The difference between an attribute's measured or true value and its nominal value.
Direct Measurements	Measurements in which a measuring attribute X directly measures the value of a subject attribute Y (i.e., X measures Y). In direct measurements, the value of the quantity of interest is obtained directly by measurement and is not determined by computing its value from the measurement of other variables or quantities.
Display Resolution	The smallest distinguishable difference between indications of a displayed value.
Distribution Function	See Cumulative Distribution Function.
Distribution Variance	The mean square dispersion of a random variable about its mean value. See also Variance.
Drift	A change in output over a period of time that is unrelated to input. Can be due aging, temperature effects, usage stress, etc.
Effective Degrees of Freedom	The degrees of freedom for a Type B estimate or a combined uncertainty estimate.
End-of-Period (EOP)	The end of the calibration or test interval.
End Item	A system, instrument or hardware component with operational performance specifications.

Term	Definition
EOP Reliability	The in-tolerance probability for an attribute at the end of its calibration or test interval.
Equipment Parameter	A specified aspect, feature or performance characteristic of a measuring device or artifact. Synonymous with attribute.
ETS	Equipment Tolerancing System. A methodology for adjusting test and calibration support parameters to minimize the total costs associated with the testing and deployment of end items.
Error	The arithmetic difference between a measured or indicated value and the true value.
Error Component	The total error in the measured or assumed value of a component variable of a multivariate measurement, e.g., the error in the measurement of distance or time in the determination of the velocity of a moving object.
Error Distribution	A probability distribution that describes the relative frequency of occurrence of values of a measurement error.
Error Equation	An expression that defines the total error in the value of a quantity in terms of all relevant process or component errors.
Error Limits	Bounding values that are expected to contain the error from a given source with some specified level of probability or confidence.
Error Model	See Error Equation.
Error Source	A parameter, variable or constant that can contribute error to the determination of the value of a quantity.
Error Source Coefficient	See Sensitivity Coefficient.
Error Source Correlation	See Correlation Analysis.
Error Source Uncertainty	The uncertainty in a given error source.
Estimated True Value	The value of a quantity obtained by Bayesian analysis.
Expanded Uncertainty	A multiple of the standard uncertainty reflecting either a specified confidence level or arbitrary coverage factor.
False Accept Risk	(1) The probability of an attribute being out-of-tolerance and falsely accepted by conformance testing as being in-tolerance. (2) The probability that attributes accepted by conformance testing are out-of-tolerance. The former is called unconditional false accept risk (UFAR) and the latter is called conditional false accept risk (CFAR).
False Reject Risk	The probability of an attribute being in-tolerance and falsely rejected by conformance testing as being out-of-tolerance.

Term	Definition
Heuristic Estimate	An estimate resulting from accumulated experience and/or technical knowledge concerning the uncertainty of an error source.
Histogram	See Sample Histogram.
Hysteresis	The lagging of an effect behind its cause, as when the change in magnetism of a body lags behind changes in an applied magnetic field.
Hysteresis Error	The residual part of a signal in a sampling event left over from the previous sampling event. In equipment specifications, the maximum separation due to hysteresis between upscale-going and downscale-going indications of a measured value taken after transients have decayed.
Independent Error Sources	Error sources that are statistically independent. See Statistical Independence.
Instrument	A device for measuring or producing the value of an observable quantity.
In-tolerance	In conformance with tolerance limits.
In-tolerance Probability	The probability that an MTE attribute value or the error in the value is contained within its specified tolerance limits.
Kurtosis	A measure of the "peakedness" of a distribution. For example, normal distributions have a peakedness value of three.
Least Significant Bit (LSB)	The smallest analog signal value that can be represented with an n-bit code. LSB is defined as $A/2^n$, where A is the amplitude of the analog signal.
Level of Confidence	See Confidence Level.
Mean Deviation	The difference between a sample mean value and a nominal value.
Mean Square Error	See Variance.
Mean Value	*Sample Mean*: The average value of a measurement sample. *Population Mean*: The expectation value for measurements sampled from a population.
Mean Value Correction	The correction or adjustment of the computed mean value for an offset due to attribute bias and/or environmental factors.
Measurement Decision Risk	The probability of erroneously accepting or rejecting an attribute based on the measurement result(s) of conformance testing.
Measurand	The particular quantity subject to measurement. (Taken from ISO GUM Annex B, Section B.2.9)
Measurement Error	The difference between the measured value of a quantity and its true value.

Term	Definition
Measurement Process Errors	Errors resulting from the measurement process (e.g., reference attribute bias, random error, resolution error, operator bias, environmental factors, …).
Measurement Process Uncertainty	The uncertainty in a measurement process error. The standard deviation of the probability distribution of a measurement process error.
Measurement Reliability	(1) Attribute Measurement Reliability: The probability that an attribute is in-tolerance. (2) Item Measurement Reliability: The probability that all attributes of an item are in-tolerance.
Measurement Reference	See Reference Standard.
Measurement Uncertainty	The lack of knowledge of the sign and magnitude of measurement error.
Measurement Units	The units, such as volts, millivolts, etc., in which a measurement or measurement error is expressed.
Measuring Device	See Measuring and Test Equipment.
Measuring and Test Equipment (MTE)	A system or device capable of being used to measure the value of a quantity or test for conformance to specifications.
Measuring Attribute	The attribute of a device that is used to obtain information that quantifies the value of a subject or unit under test attribute.
Median Value	(1) The value that divides an ordered sample of data in two equal portions. (2) The value for which the distribution function of a random variable is equal to one-half. (3) A point of discontinuity such that the distribution function immediately below the point is less than one-half and the distribution function immediately above the point is greater than one-half.
Mode Value	The value of an attribute most often encountered or measured. Sometimes synonymous with the nominal value or design value of an attribute.
Module Error Sources	Sources of error that accompany the conversion of module input to module output.
Module Input Uncertainty	The uncertainty in a module's input error expressed as the uncertainty in the output of preceding module.
Module Output Equation	The equation that expresses the output from a module in terms of its input. The equation is characterized by parameters that represent the physical processes that participate in the conversion of module input to module output.
Module Output Uncertainty	The total combined uncertainty in the output of a given module of a measurement system.

Term	Definition
Multivariate Measurements	Measurements in which the value of a subject attribute is a computed quantity based on measurements of two or more variables.
Nominal Accuracy Ratio	For two-sided tolerance limits symmetric about a nominal value, the ratio of the span of UUT attribute tolerance limits to the span of the tolerance limits of a test or calibration reference attribute.
Nominal Value	The designated or published value of an attribute. It may also sometimes refer to the mode value of an attribute's distribution.
Operator Bias	The systematic error due to the perception or influence of a human operator or other agency.
Parameter	A characteristic of a device, process or function. See also Equipment Parameter.
Parameter Value Equation	The system equation for a multivariate measurement.
Population	The total set of possible values for a random variable.
Population Mean	The expectation value of a random variable described by a probability distribution.
Precision	The number of places to the right the decimal point in which the value of a quantity can be expressed. Although higher precision does not necessarily mean higher accuracy, the lack of precision in a measurement is a source of measurement error.
Probability	The likelihood of the occurrence of a specific event or value from a population of events or values.
Probability Density Function (pdf)	A mathematical function that describes the relative frequency of occurrence of the values of a random variable.
Quantization	The sub-division of the range of a reading into a finite number of steps, not necessary equal, each of which is assigned a value. Particularly applicable to analog to digital and digital to analog conversion processes.
Quantization Error	Error due to the granularity of resolution in quantizing a sampled signal. Contained within +/- 1/2 LSB (least significant bit).
Random Error	See Repeatability.
Range	An interval of values for which specified tolerances apply. In a calibration or test procedure, a setting or designation for the measurements of a set of specific points.
Readout Device	A device that converts a signal to a series of numbers on a digital display, the position of a pointer on a meter scale, tracing on recorder paper or graphic display on a screen.

Term	Definition
Reference Standard	An artifact used as a measurement reference whose value and uncertainty have been determined by calibration and documented.
Reliability Model	A mathematical function relating the in-tolerance probability of an attribute or instrument and the time elapsed since testing or calibration. Used to project uncertainty growth over time.
Repeatability	The error that manifests itself in the variation of the results of successive measurements of the value of a quantity carried out under the same measurement conditions and procedure during a measurement session. Also referred to as Random Error.
Reproducibility	The closeness of the agreement between the results of measurements of the value of an attribute carried out under different measurement conditions. The differences may include: principle of measurement, method of measurement, observer, measuring instrument(s), reference standard, location, conditions of use, time.
Resolution	The smallest discernible value indicated by a reference or subject attribute.
Resolution Error	The error due to the finiteness of the precision of a measurement.
Sample	A collection of values drawn from a population from which inferences about the population are made.
Sample Histogram	A bar chart showing the relative frequency of occurrence of sampled values.
Sample Mean	The arithmetic average of sampled values.
Sample Size	The number of values that comprise a sample.
Sensitivity	The ratio between a change in the electrical output signal to a small change in the physical input of a sensor or transducer. The derivative of the transfer function with respect to the physical input.
Sensitivity Coefficient	A coefficient that weights the contribution of a given error component to the total measurement error.
Sensor	Any of various devices designed to detect, measure or record physical phenomena.
Skewness	A measure of the asymmetry of a probability distribution. A symmetric distribution has zero skewness.
Specification	A numerical value or range of values that bound the performance of an MTE attribute.
Stability	The ability of a measuring device to give constant output for a constant input over a period of time.

Term	Definition
Standard Deviation	The square root of the variance of a sample or population of values. A quantity that represents the spread of values about a mean value. In statistics, the second moment of a distribution.
Standard Uncertainty	The standard deviation of an error distribution.
Statistical Independence	A property of two or more random variables such that their joint probability density function is the product of their individual probability density functions. Two error sources are statistically independent if one does not exert an influence of the other or if both are not consistently influenced by a common agency. See also Independent Error Sources.
Stress Response Error	The error or bias in an attribute value induced by response to applied stress.
Student's t-statistic	Typically expressed as $t_{\alpha,v}$. Denotes the value for which the distribution function for a t-distribution with v degrees of freedom is equal to $1 - \alpha$. A multiplier used to express an error limit or expanded uncertainty as a multiple of a standard uncertainty.
Subject Attribute	An attribute whose value we seek to obtain from a measurement or set of measurements.
Symmetric Distribution	A probability distribution of random variables that are equally likely to be found above or below a mean value.
System Equation	A mathematical expression that defines the value of a quantity in terms of its constituent variables or components. Also referred to as a parameter value equation.
System Module	An intermediate stage of a system that transforms an input quantity into an output quantity according to a module output equation.
System Output Uncertainty	The total uncertainty in the output of a measurement system
t Distribution	A symmetric, continuous distribution characterized by the degrees of freedom parameter. Used to compute confidence limits for normally distributed variables whose estimated standard deviation is based on finite degrees of freedom. Also referred to as the Student's t distribution.
Test Accuracy Ratio	An alternative label for Test Uncertainty Ratio.
Test Uncertainty Ratio	The ratio of the span of the tolerance limits of a UUT attribute and two times the 95% expanded measurement uncertainty of a conformance testing measuring process.
Tolerance Limits	Typically, engineering tolerances that define the maximum and minimum values for a product to work correctly. These tolerances bound a region that contains a certain proportion of the total population with a specified probability or confidence.

Term	Definition
Total Uncertainty	The standard deviation of the probability distribution of the total combined error in the value of a quantity obtained by measurement.
Total System Uncertainty	See System Output Uncertainty.
True Value	The value that would be obtained by a perfect measurement. True values are by nature indeterminate.
Type A Estimates	Uncertainty estimates obtained by statistical analysis of a sample of data.
Type B Estimates	Uncertainty estimates obtained by heuristic means in the absence of a sample of data.
Uncertainty	See Standard Uncertainty.
Uncertainty Component	The uncertainty in an error component.
Uncertainty in the Mean Value	The standard deviation of the distribution of mean values obtained from multiple sample sets of measurements of values of a given quantity. Estimated by the standard deviation of a single sample divided by the square root of the sample size. The distribution of mean values is called the *sampling distribution*.
Uncertainty Ratio	The ratio of the UUT attribute bias standard uncertainty to the standard uncertainty of the test or calibration process at the time of testing or calibration.
Unconditional False Accept Risk (*UFAR*)	The probability that an equipment attribute will be out-of-tolerance and accepted as in-tolerance.
Uncertainty Growth	The increase in the uncertainty in the value or bias of an attribute over the time elapsed since measurement.
Unit Under Test	A device featuring the subject attribute.
Variance	(1) Population: The expectation value for the square of the difference between the value of a variable and the population mean. (2) Sample: A measure of the spread of a sample equal to the sum of the squared observed deviations from the sample mean divided by the degrees of freedom for the sample. Also referred to as the mean square error.
Within Sample Sigma	An indicator of the variation within samples.

1.5 Acronyms

A/D	Analog to Digital
ANSI	American National Standards Institute
AOP	Average-Over-Period
AR	Accuracy Ratio
ATE	Automated Test Equipment

BIPM	International Bureau of Weights and Measures (Bureau International des Poids et Mesures)
BOP	Beginning-Of-Period
CFAR	Conditional False Accept Risk. Sometimes called CPFA
CGPM	General Conference on Weights and Measures (Conference General des Poids et Mesures)
CIPM	International Conference of Weights and Measures (Conference Internationale des Poids et Mesures)
CPFA	Conditional Probability for a False Accept. See CFAR.
DVM	Digital Voltmeter
EOP	End-of-Period
ESS	Pure Error Sum of Squares
FAR	False Accept Risk
FRR	False Reject Risk
FS	Full Scale
GUM	Guide to the Expression of Uncertainty in Measurement
IS0	International Organization for Standardization (Organisation Internationale de Normalisation)
LCL	Lower Confidence Limit
LSS	Lack of Fit Sum of Squares
MAP	Measurement Assurance Program
MDR	Measurement Decision Risk
MLE	Maximum-Likelihood-Estimate
MQA	Measurement Quality Assurance
MTE	Measuring and Test Equipment
NASA	National Aeronautics and Space Administration
NBS	National Bureau of Standards (now NIST)
NHB	NASA Handbook
NIST	National Institute of Standards and Technology (was NBS)
pdf	Probability Density Function
PFA	Probability for a False Accept. See UFAR.
PFR	Probability for a False Reject. See FRR.
PRT	Platinum Resistance Thermometer
QA	Quality Assurance
RSS	Root-Sum-Square
SI	International System of Units (Système International d'Unités)
SMPC	Statistical Measurement Process Control
SPC	Statistical Process Control
TAR	Test Accuracy Ratio
TUR	Test Uncertainty Ratio
UCL	Upper Confidence Limit
UFAR	Unconditional False Accept Risk. Sometimes called PFA.
UUT	Unit Under Test
VIM	International Vocabulary of Basic and General Terms in Metrology (Vocabulaire International des Termes Fondamentaux et Généraux de Métrologie)

Chapter 2: Uncertainty Analysis Overview

Estimating measurement decision risk begins with the estimation of the uncertainty in the measurement of the quantity of interest and of the uncertainty in the bias of this quantity prior to measurement. For this reason, a brief discussion of measurement uncertainty analysis principles and methods is provided in this chapter. In addition to covering principles and methods, a description of four test or calibration scenarios is presented. These scenarios provide prescriptions for combining uncertainties that ensure measurement decision risk analyses are compatible with specific measurement processes.

The uncertainty analysis overview given in this chapter is augmented by a more detailed discussion presented in Appendix A. For a full discussion and examination of the subject of uncertainty analysis, the reader is referred to more in-depth measurement uncertainty analysis references [26, 27 and Annex 3].

2.1 Uncertainty Analysis and Risk Management

Uncertainty analysis is vital to the effective management of modern technology. Cogent uncertainty estimates provide statistics that can be used in making technology management decisions. Uncertainty estimates support activities that range from assessing the compatibility of parts to computing the risks involved in making decisions based on measurement results. As is shown in Appendix B, uncertainty analysis is essential for developing important calibration quality metrics.

2.1.1 Measurement Decision Risk

Measurement uncertainty estimates are important variables in computing false accept and false reject risks in calibration and testing. Knowing these risks can lead to the avoidance of placing nonconforming attributes in use and can reduce costs associated with reworking rejected attributes. Using uncertainty estimates to control risks can also come into play when determining acceptance limits for testing or calibration.

2.1.2 Parts Conformance

Cannons and Cannonballs

It may be surmised that a primary objective of recognized standards, such as ANSI/NCSL Z540.3-2006 [33] and ISO/IEC 17025 [A-3] is to assure that parts manufactured by one company will be compatible with parts manufactured by another company. To illustrate, consider the hypothetical case of ensuring the utility of a particular type of cannon. This utility can be expressed primarily in terms of reliability, portability and range. With respect to range, the variables of interest are muzzle velocity, cannonball mass and cannonball aerodynamics. Obviously, muzzle velocity is dependent in large part on the difference between the inner diameter of the cannon and the diameter of the cannonballs.

Clearly, if Company X manufactures the cannon and Company Y manufactures the cannonballs, their measurements must be in close enough agreement to ensure that diameter differences are not too large or too small. This involves comparing relative specifications and taking into account uncertainties in measurement.

2.1.3   Statistical Tolerancing

It is often desirable to manufacture items whose attributes conform to specifications with a certain level of confidence. Such specifications can be developed from estimates of the uncertainties (variabilities) of attribute values of manufactured/tested attributes and associated degrees of freedom. It should be stated that, if uncertainty estimates are overly conservative, computed confidence levels will likely not be useful, and tolerance limits will be specified wider than they should be. This can have an undesirable impact on equipment selection and operational equipment support costs.

2.2   Uncertainty Analysis Fundamentals

We now present a short discussion of the elements of uncertainty estimation. After some preliminary remarks, we will identify types of estimates and outline the estimation process for each.

2.2.1   Preliminaries

In estimating uncertainty, we keep in mind a few basic concepts. Chief among these are the following:

- All measurements are accompanied by error.

- Measurement errors and attribute biases are <u>random variables</u>. This means that whenever we make measurements, the sign and magnitude of errors in measurement results are unknown and vary unpredictably.

- Errors follow <u>probability distributions</u>. The way that measurement errors vary can be described statistically. In statistical descriptions, errors are said to be distributed in such a way that the sign and magnitude of a given error has associated with it a probability of occurrence. The key words in this view of measurement errors are

- **Population** - All the values that a random variable can attain.

- **Distribution** - A functional relationship between the value of a random variable and its probability of occurrence.

- The lack of knowledge of the value of a measurement error is called <u>measurement uncertainty</u>.

- The uncertainty in the value of a measurement error is the <u>standard deviation</u> of the measurement error distribution.

From the above, we see that uncertainty analysis attempts to quantify the probability distributions of measurement errors by estimating the standard deviations of their distributions.

2.2.2   The Basic Error Model

The basic error model applies to the measurement x_{meas} of a quantity x_{true}

$$x_{meas} = x_{true} + \varepsilon_{meas},$$

where ε_{meas} is the measurement error. If x_{meas} is measured directly,

$$\varepsilon_{meas} = \varepsilon_{bias} + \varepsilon_{repeat} + \varepsilon_{resolution} + \varepsilon_{operator} + \varepsilon_{environment} + \cdots$$

Each contributing error in a direct measurement is called an *error source*. If x_{meas} is obtained by multivariate measurement of m components [A3], then

$$\varepsilon_{meas} = \sum_{i=1}^{m} a_i \varepsilon_i ,$$

where the constants a_i are sensitivity coefficients. Each error ε_i is called an *error component*. Error components may be comprised of one or more error sources, each arising from a direct measurements

2.2.3   Measurement Error Sources

Various sources of error may contribute to the total combined measurement error. Most commonly encountered are the following:

Measurement Bias

A systematic difference between the value of a UUT attribute measured with a measurement reference and the attribute's true value.

Random Error

Error which manifests itself in differences between measurements within a measurement sample.

Resolution Error

The difference between a measured (sensed) value and the value indicated by a measuring device.

Digital Sampling Error

Error due to the granularity of digital representations of analog values.

Computation Error

Error due to computational round-off and other errors due to extrapolation, interpolation, curve fitting, etc.

Operator Bias

Error due to a persistent bias in operator perception and/or technique.

Stress Response Error

Error caused by response to stress following measurement.

Environmental/Ancillary Error

Error caused by environmental effects and/or biases or fluctuations in ancillary equipment.

2.2.4   Error and Uncertainty

2.2.4.1   Axioms

Three axioms important for understanding measurement uncertainty and performing uncertainty analysis are given below.

Axiom 1:

Measurement errors follow probability distributions.[4]

Axiom 2:

The uncertainty in a measurement of a variable is the square root of the variable's distribution variance or "mean square error."

$$u_x = \sqrt{\text{var}(x_{meas})}$$

Axiom 3:

The uncertainty in a measurement is the uncertainty in the measurement error.

$$\begin{aligned}
u_x &= \sqrt{\text{var}(x_{meas})} \\
&= \sqrt{\text{var}(x_{true} + \varepsilon_{meas})} \\
&= \sqrt{\text{var}(\varepsilon_{meas})} \\
&= u_{\varepsilon_x} .
\end{aligned}$$

2.2.4.2   Variance Addition Rule

The fundamental tool for combining uncertainties due to error sources or components is the variance operator. The recipe for combining uncertainties using this operator is the variance addition rule. It can be expressed in two ways. For example, if a function $z = ax + by$, then

Covariance Version

$$\begin{aligned}
u_z^2 = \text{var}(z) &= \text{var}(ax + by) \\
&= a^2 \text{var}(x) + b^2 \text{var}(y) + 2ab \text{cov}(x, y) \\
&= a^2 \text{var}(\varepsilon_x) + b^2 \text{var}(\varepsilon_y) + 2ab \text{cov}(\varepsilon_x, \varepsilon_y) \\
&= a^2 u_{\varepsilon_x}^2 + b^2 u_{\varepsilon_y}^2 + 2ab \text{cov}(\varepsilon_x, \varepsilon_y) \\
&= u_{\varepsilon_z}^2
\end{aligned}$$

---

[4] See Section A.2.2.2.

Correlation Version

$$
\begin{aligned}
u_z^2 &= \text{var}(z) = \text{var}(ax + by) \\
&= a^2 \,\text{var}(x) + b^2 \,\text{var}(y) + 2ab\rho_{x,y}\sqrt{\text{var}(x)\,\text{var}(y)} \\
&= a^2 \,\text{var}(\varepsilon_x) + b^2 \,\text{var}(\varepsilon_y) + 2ab\rho_{x,y}\sqrt{\text{var}(\varepsilon_x)\,\text{var}(\varepsilon_y)} \\
&= a^2 u_{\varepsilon_x}^2 + b^2 u_{\varepsilon_y}^2 + 2ab\rho_{x,y} u_{\varepsilon_x} u_{\varepsilon_y} \\
&= u_{\varepsilon_z}^2
\end{aligned}
$$

$$x = x_{true} + \varepsilon_x$$

$$\varepsilon_x = \varepsilon_b + \varepsilon_{ran} + \varepsilon_{res} + \varepsilon_{op} + \varepsilon_{env} + \cdots$$

$$
\begin{aligned}
u_x^2 &= \text{var}(x) = \text{var}(x_{true} + \varepsilon_x) \\
&= \text{var}(\varepsilon_x) = u_{\varepsilon_x}^2 = \text{var}(\varepsilon_b + \varepsilon_{ran} + \varepsilon_{res} + \varepsilon_{op} + \varepsilon_{env} + \cdots) \\
&= u_{\varepsilon_b}^2 + u_{\varepsilon_{ran}}^2 + u_{\varepsilon_{res}}^2 + u_{\varepsilon_{op}}^2 + u_{\varepsilon_{env}}^2 + \cdots \\
&\quad + 2\rho_{\varepsilon_b\varepsilon_{ran}} u_{\varepsilon_b} u_{\varepsilon_{ran}} + 2\rho_{\varepsilon_b\varepsilon_{res}} u_{\varepsilon_b} u_{\varepsilon_{res}} + \cdots
\end{aligned}
$$

Figure 2-1. The Variance Addition Rule for Measurement Errors

Shown is the correlation version. Note: For direct measurements, correlations between process errors are usually zero.

2.2.5 Procedures for Obtaining an Uncertainty Estimate for an Error Source

2.2.5.1 Type A Estimates

In making a Type A estimate and using it to construct confidence limits, we apply the following procedure taken from the GUM and elsewhere:

1. Take a random sample of size n representative of the population of interest. The larger the sample size, the better. In many cases, a sample size less than six is not sufficient.

2. Compute a sample mean

$$\bar{x} = \frac{1}{n}(x_1 + x_2 + \cdots + x_n) = \frac{1}{n}\sum_{i=1}^{n} x_i$$

3. Compute a sample standard deviation u_x

$$s_x = \sqrt{\frac{1}{n-1}\sum_{i=1}^{n}(x_i - \bar{x})^2}\ ,$$

   where the x_i, $i = 1, 2, \cdots, n$ comprise a sample of n measured values.

4. Assume an underlying distribution, e.g., normal.

5. Develop a coverage factor equal to a *t*-statistic based on the degrees of freedom $v = n - 1$ associated with the sample standard deviation and a desired level of confidence p. Multiply the sample standard deviation by the coverage factor (*t*-statistic) to obtain confidence limits $\pm L$:

6. If reporting the sample mean value or basing decision on this value, divide the standard deviation s_x by the square root of the sample size n.

Degrees of Freedom

$$v = n - 1$$

Significance Value α

$$\alpha = 1 - p$$

Single-Measurement Repeatability Confidence Limits

$$L = \pm t_{\alpha/2,v} s_x$$

Mean Value Repeatability Confidence Limits:

$$L = \pm t_{\alpha/2,v} \frac{s_x}{\sqrt{n}}$$

2.2.5.2   Type B Estimates

In making a Type B estimate, we reverse the process. The procedure is

1. Develop a set of error containment limits $\pm L$.
2. Estimate a containment probability p.
3. Estimate the Type B degrees of freedom[5]
4. Assume an underlying distribution, e.g., normal.
5. Compute a coverage factor, t, based on the containment probability and degrees of freedom.
6. Compute the standard uncertainty for the quantity of interest (e.g., attribute bias) by dividing the containment limit by the coverage factor: $u = L / t$.

2.2.6   Degrees of Freedom for Combined Estimates

The degrees of freedom for a total combined uncertainty u_T, made up of k uncertainties u_i for s-independent errors ε_i, $i = 1,2,\cdots,k$, is given by the Welch-Satterthwaite relation [26, 41, 42][6]

$$v_T = \frac{u_T^4}{\displaystyle\sum_{i=1}^{k} \frac{u_i^4}{v_i}} .$$

If errors ε_i and ε_j are correlated with correlation coefficient ρ_{ij}, $i,j = 1,2,\cdots,k$, then the above relation may be only approximately valid. A version of the Welch-Satterthwaite relation that has been proposed for use with correlated errors is

---

[5] See reference [26] or [27] for details regarding computing Type B degrees of freedom.

[6] If, in the equation for u_T, an uncertainty component u_i is multiplied by a sensitivity coefficient c_i, the u_i term in the Welch-Satterthwaite relation is replaced by the term $c_i u_i$.

$$v = \frac{u_T^4}{\sum_{i=1}^{k}\frac{u_i^4}{v_i} + 2\sum_{i=1}^{k-1}\sum_{j=i+1}^{k}\rho_{ij}^2\sigma^2(u_i)\sigma^2(u_j)}$$

where

$$u_T = \sum_{i=1}^{n}u_i^2 + 2\sum_{i=1}^{n-1}\sum_{j=i+1}^{n}\rho_{ij}u_i u_j .$$

This version of the relation has been circulated to selected members of the measurement science community for review and comment. Its derivation is given in Appendix K.

2.2.7   Expanded Uncertainty

The expanded uncertainty is a limit obtained by multiplying an uncertainty estimate by a specified **coverage factor**. To develop expanded uncertainties that can be used to make cogent decisions for technology management, it is desirable to relate the coverage factor to a confidence level. If this is done, then the expanded uncertainty serves as a **confidence limit** that can be said to bound errors with a stated degree of confidence.

Suppose, for example, that a mean value $\bar{x}$ and an uncertainty estimate u have been obtained for a variable x, along with a degrees of freedom estimate v. If it is desired to establish $\pm$ confidence limits around $\bar{x}$ that bound errors in x with some probability p, then the expanded uncertainty is given by[7]

$$L = t_{\alpha/2,v}u ,$$

where the coverage factor $t_{\alpha/2,v}$ is the familiar Student's t-statistic and $\alpha = 1 - p$. In using the expanded uncertainty to indicate the confidence in the estimate $\bar{x}$, we would say that the value of x is estimated as $\bar{x} \pm L$ with $p \times 100\%$ confidence.

Some institutional procedures employ a fixed, arbitrary number to use as a coverage factor. The argument for this practice is based on the assertion that the degrees of freedom for Type B estimates cannot be rigorously established. Accordingly, it makes little sense to attempt to determine a t-statistic based on a confidence level and degrees of freedom for a combined Type A/B uncertainty. Until the late '90s, this assertion had been true. Methods now exist that permit the determination of Type B degrees of freedom.[8] Given this, we are no longer limited to the practice of using a fixed number for all expanded uncertainty estimates.

2.3   Multivariate Uncertainty Analysis

Frequently, the value of a quantity of interest is obtained by measuring the values of constituent quantities or "components." An example is the measurement of velocity, obtained through measurements of time and distance. In such cases, an **error model** is required as a starting point in developing an expression for the uncertainty in the quantity of interest.

---

[7] The distribution for the population of errors in x is assumed to be normal in this example; hence the use of the t-statistic for computing confidence limits.

[8] Castrup, H., "Estimating Category B Degrees of Freedom," Measurement Science Conference, Anaheim, January 21, 2000.

2.3.1 Error Modeling

Error modeling consists of identifying the various components of error and establishing their relationship to one another. The guiding expression for this process is the **system equation**.

2.3.1.1 The System Equation

The system equation is the expression for the variable being sought in terms of its measurable components. Establishing the system equation is often the most difficult part of the process. If the system equation can be determined, then uncertainty analysis becomes almost trivial.

For purposes of illustration, we consider a two-component variable. The expressions that ensue can easily be extended to cases with arbitrary numbers of components.

Let the component variables of the system equation be labeled x and y. Then, if the variable of interest, labeled z, is expressed as a function of x and y, we write

$$z = z(x, y).$$

2.3.1.2 Error Components

The measurement error or bias in each variable in the system equation is an error component. The contribution of each error component ε_x and ε_y to the error ε_z in the variable z is expressed in the error model

$$\varepsilon_z = \left(\frac{\partial z}{\partial x}\right)\varepsilon_x + \left(\frac{\partial z}{\partial y}\right)\varepsilon_y + O(2),$$

where $O(2)$ indicates terms to second order and higher in the error variables. Higher order terms are usually negligible and are dropped from the expression to yield

$$\varepsilon_z \cong \left(\frac{\partial z}{\partial x}\right)\varepsilon_x + \left(\frac{\partial z}{\partial y}\right)\varepsilon_y .$$

2.3.2 Computing System Uncertainty

Using Axioms 2 and 3, together with the variance addition rule, gives

$$u_z = \sqrt{c_x^2 u_x^2 + c_y^2 u_y^2 + 2 c_x c_y \rho(\varepsilon_x, \varepsilon_y) u_x u_y} ,$$

where the coefficients c_x and c_y are

$$c_x = \left(\frac{\partial z}{\partial x}\right), \qquad c_y = \left(\frac{\partial z}{\partial y}\right),$$

and the uncertainties are

$$u_x = \sqrt{\text{var}(\varepsilon_x)}, \qquad u_y = \sqrt{\text{var}(\varepsilon_y)} .$$

The coefficient $\rho(\varepsilon_x, \varepsilon_y)$ is the correlation coefficient for the component errors ε_x and ε_y.

2.3.3 Process Uncertainties

Various sources of error contribute to each error component of a measurement. Most commonly encountered are those described earlier in Section 2.2.3:

- Measurement Bias
- Random Error
- Resolution Error
- Digital Sampling Error
- Computation Error
- Operator Bias
- Stress Response Error
- Environmental/Ancillary Error

Labeling each of the relevant error sources with a number designator, we can write the error in the measurement of the component x as

$$\varepsilon_x = \varepsilon_{x1} + \varepsilon_{x2} + \mathrm{L} \ \varepsilon_{xn},$$

and, since correlations between error sources within an error component are generally zero, the uncertainty in the measurement of x becomes a simple RSS combination

$$u_x^2 = u_{x1}^2 + u_{x2}^2 + \mathrm{L} + u_{xn}^2,$$

Similarly, the uncertainty in the measurement of y is given by

$$u_y^2 = u_{y1}^2 + u_{y2}^2 + \mathrm{L} + u_{yn}^2.$$

2.3.4 Cross-Correlations

The final topic of this section deals with cross-correlations between error sources of different error components. Cross-correlations occur when different components of a variable are measured using the same device, in the same environment, by the same operator or in some other way that would lead us to suspect that the measurement errors in the two components might be correlated.

If the cross-correlation between the *ith* and the *jth* process errors of the measured variables x and y is denoted by $\rho(\varepsilon_{xi}, \varepsilon_{yj})$, then the correlation coefficient between x and y is given by

$$\rho(\varepsilon_x, \varepsilon_y) = \frac{1}{u_x u_y} \sum_{i=1}^{n_i} \sum_{j=1}^{n_j} \rho(\varepsilon_{xi}, \varepsilon_{yj}) u_{xi} u_{yj}.$$

2.4 Uncertainty Analysis Scenarios

As stated at the beginning of this chapter, Appendix A describes four calibration scenarios for estimating and combining uncertainties. The four scenarios are the following:

1. The measurement reference (MTE) measures the value of an attribute of the unit under test (UUT) that provides an output or stimulus.

2. The UUT measures the value of a reference attribute of the MTE that provides an output or stimulus.

3. The UUT and MTE each provide an output or stimulus" for comparison using a bias cancellation comparator.

4. The UUT and MTE both measure the value of an attribute of a common device or artifact that provides an output or stimulus.

The information obtained in each scenario includes an observed value, referred to as a "measurement result" or "calibration result," and an estimated uncertainty in the calibration error. For each scenario, a measurement equation is given that is applicable to the manner in which calibrations are performed and calibration results are recorded or interpreted.

For each of the four scenarios, the calibration result is expressed as

$$\delta = e_{UUT,b} + \varepsilon_{cal} \,,$$

where $e_{UUT,b}$ is the bias of the unit-under-test (UUT) at the time of calibration, δ is the measurement (estimate) of $e_{UUT,b}$ and ε_{cal} is the total calibration error in the estimate. The calibration uncertainty is given by

$$u_{cal} = \sqrt{\operatorname{var}(\varepsilon_{cal})} \,,$$

where $\operatorname{var}(\varepsilon_{cal})$ is the variance in the probability distribution of ε_{cal}.

2.5   Interpreting and Applying Equipment Specifications

Equipment specifications are an important element of testing, calibration and other measurement processes. They are used for the selection of MTE or for establishing equipment substitutions for a given measurement application. In addition, manufacturer specified tolerances are used to compute test uncertainty ratios and estimate bias uncertainties.

The subject of interpreting and applying equipment specifications in uncertainty analysis and risk analysis deserves a full and unabridged discussion that is beyond the scope of this Handbook Annex. This discussion is provided in reference [35].

2.6   Uncertainty Analysis Examples

Examples of uncertainty analyses for illustrating the concepts and procedures outlined in this chapter, as well as examples for each scenario summarized above and discussed in Appendix A are provided in [26] and [27].

Chapter 3: Measurement Decision Risk Analysis Basics

Measurement decision risk is defined in terms of probabilities of the occurrence of events relating to testing and calibration. This chapter describes these events and develops the probability relations that are used to compute measurement decision risk. Basic probability concepts are covered in Section 3.1.

In computing measurement decision risk, we may resort to test limits or adjustment limits that differ from attribute tolerance limits. These limits are referred to as "guardband limits." The probability expressions needed to incorporate guardband limits in the estimation of risk will be developed in Chapter 5.

Finally, we will take a different view of false accept risk in considering the probability that a given specific attribute will be accepted as out-of-tolerance under the alternative renewal policies of (1) not adjusting to nominal and (2) adjusting to nominal. To develop the necessary probability expressions, Bayesian analysis will be employed [28-30].

3.1   Preliminaries

Before establishing the probability expressions for measurement decision risk, we will discuss probability functions in general terms. First, we will define probability. Next, the concepts of joint and conditional probability will be discussed. Finally, we will examine a relation that is particularly useful in risk analysis. This relation is referred to as Bayes' theorem [34].

3.1.1   Definition of Probability

There are two basic ways to define probability. One stems from fundamental considerations and the other links the definition of probability to empirical evidence.

In the first definition, we enumerate all the possible events or *outcomes* that are possible within a given context. For example, the context may be flipping a coin. In this case the possible outcomes are "heads" and "tails." Another example is throwing a die. There are six possible outcomes, all equally probable.

In the second definition, we gather data for a particular quantity and converge to the probability of obtaining specific values. For example, suppose the quantity is the net distance a rubber ball bounces from ground zero if dropped from a particular height. If we drop the ball from the height $N = 100$ times and obtain a value of $N_E = 6$ for a net distance of 2.2 meters, we estimate the probability of obtaining 2.2 meters to be

$$P(E) \cong N_E / N = 0.06.$$

This probability estimate becomes an actual probability of occurrence as the number $N \rightarrow \infty$:

$$P(E) = \lim_{N \rightarrow \infty} N_E / N. \tag{3-1}$$

This relation is called the law of large numbers.[9]

3.1.2   Joint Probability

In risk analysis, we are often interested in the probability of two events occurring simultaneously. For example, we might want to know the probability that a UUT attribute is both in-tolerance and perceived as being in-tolerance. If we represent the event of an in-tolerance attribute as E_1 and the event of observing the attribute to be in-tolerance as E_2, then the joint probability for occurrence of E_1 and E_2 is written

$$P(E_1 \text{ and } E_2) = P(E_1, E_2). \tag{3-2}$$

3.1.2.1   Statistical Independence

If the occurrence of event E_1 and the occurrence of event E_2 bear no relationship to one another, they are called statistically independent. For example, E_1 may represent the outcome that an individual selected at random from within a group is 30 years old and E_2 may represent the event that his shoe size is 11.[10]

It can be shown that, for such statistically independent events,

$$P(E_1, E_2) = P(E_1)P(E_2). \tag{3-3}$$

Another important result derives from the probability that event E_1 will occur or event E_2 will occur. The appropriate relation is

$$P(E_1 \text{ or } E_2) = P(E_1) + P(E_2) - P(E_1, E_2). \tag{3-4}$$

Combining this expression with the preceding one gives the relation for cases where E_1 and E_2 are independent

$$P(E_1 \text{ or } E_2) = P(E_1) + P(E_2) - P(E_1)P(E_2). \tag{3-5}$$

As an example, consider the probability that the bias in a reference attribute e_{bias} is less than 2 mV, and the random error (repeatability) in the measurement process e_{ran} is less than 1 mV. Since attribute bias is independent of random error, we can write

$$P(e_{bias} < 2\,mV \text{ or } e_{ran} < 1\,mV) = P(e_{bias} < 2\,mV) + P(e < 1\,mV) \\ -P(e_{bias} < 2\,mV)P(e_{ran} < 1\,mV). \tag{3-6}$$

3.1.2.2   Mutually Exclusive Events

On occasion, events are mutually exclusive. That is, they cannot occur together. A popular example is the tossing of a coin. Either heads will occur or tails will occur. They obviously cannot occur simultaneously. This means $P(E_1, E_2) = 0$, and

$$P(E_1 \text{ or } E_2) = P(E_1) + P(E_2). \tag{3-7}$$

---

[9] Jacob Bernoulli first described the law of large numbers as, "so simple that even the stupidest of men instinctively know it is true."

[10] Examples of independent events are found in Appendix A for errors that are uncorrelated.

3.1.3   Conditional Probability
3.1.3.1   General

If the occurrence of E_2 is influenced by the occurrence of E_1, we say that E_1 and E_2 are conditionally related and that the probability of E_2 is conditional on the event E_1. Conditional probabilities are written

$$P(E_2 \text{ given } E_1) = P(E_2 \mid E_1). \tag{3-8}$$

It can be shown that the joint probability for E_1 and E_2 can be expressed as

$$P(E_2, E_1) = P(E_2 \mid E_1)P(E_1). \tag{3-9}$$

Equivalently, we can also write

$$P(E_1, E_2) = P(E_1 \mid E_2)P(E_2). \tag{3-10}$$

Note that, since $P(E_1, E_2) = P(E_2, E_1)$, we have

$$P(E_1 \mid E_2)P(E_2) = P(E_2 \mid E_1)P(E_1). \tag{3-11}$$

We will return to this result later.

3.1.3.2   Mutually Exclusive Events

Let E represent an outcome with k mutually exclusive possible causes $A_1, A_2, \cdots, A_k$. The probability of observing E is given by

$$
\begin{aligned}
P(E) &= P(E \mid A_1)P(A_1) + P(E \mid A_2)F(A_2) \\
&\quad + \ldots + P(E \mid A_k)P(A_k) \\
&= P(E, A_1) + P(E, A2) + \text{L} + P(E, A_k).
\end{aligned}
$$

Note that an event cannot both occur and not occur. Accordingly, these two outcomes are mutually exclusive. We write the non-occurrence of an event E by capping it with a bar. i.e.,

$$\text{Probability that } E \text{ will not occur} = P(\overline{E}),$$

and, since E and $\overline{E}$ are mutually exclusive,

$$P(E) + P(\overline{E}) = 1.$$

We use this fact to develop a special case of the rule for mutually exclusive events that is useful for risk analysis. Given that we have mutually exclusive events E_1 and E_2, we can write

$$P(E_1) = P(E_1, E_2) + P(E_1, \overline{E}_2). \tag{3-12}$$

Of course, the same applies for $P(E_2)$:

$$P(E_2) = P(E_1, E_2) + P(\overline{E}_1, E_2). \tag{3-13}$$

To illustrate, suppose that

$\quad E_1 \quad = \quad$ the event that an attribute is in-tolerance, and

$\quad E_2 \quad = \quad$ the event that the attribute is *observed* to be in-tolerance.

Then, the probability that the attribute is observed to be in-tolerance can be written

$$P(E_2) = P(E_1, E_2) + P(\bar{E}_1, E_2), \qquad (3\text{-}14)$$

and the probability that the UUT attribute is in-tolerance can be written

$$P(E_1) = P(E_1, E_2) + P(E_1, \bar{E}_2). \qquad (3\text{-}15)$$

Given these expressions, we can express the probability that the attribute will be out-of-tolerance and observed to be in-tolerance as

$$P(\bar{E}_1, E_2) = P(E_2) - P(E_1, E_2), \qquad (3\text{-}16)$$

and the probability that the attribute will be in-tolerance and observed to be out-of-tolerance as

$$P(E_1, \bar{E}_2) = P(E_1) - P(E_1, E_2). \qquad (3\text{-}17)$$

These results will be used later in defining measurement decision risk.

3.2   False Accept Risk

The "out-the-door" quality of a calibration or testing organization engaged in conformance testing can be evaluated in terms of the probability that attributes that are accepted as being in-tolerance are actually out-of-tolerance. This probability is called **false accept risk**. There are two alternatives for false accept risk; **Unconditional False Accept Risk** ($UFAR$)[11] and **Conditional False Accept Risk** ($CFAR$).

$UFAR$ is the probability that a UUT attribute will be both out-of-tolerance and perceived as being in-tolerance during testing or calibration. $CFAR$, on the other hand, is the probability that an attribute accepted by conformance testing will be out-of-tolerance.

These risks are defined in terms of probabilities in the following sections. In these definitions, we use the notation

E_L -   The event that the UUT attribute is in-tolerance.

E_A -   The event that the UUT attribute is observed to be in-tolerance.

3.2.1   Unconditional False Accept Risk

The probability that the events $\bar{E}_L$ and E_A will both occur is written

$$UFAR = P(\bar{E}_L \text{ and } E_A) = P(\bar{E}_L, E_A). \qquad (3\text{-}18)$$

Since, by Eq. (3-16), $P(\bar{E}_L, E_A) = P(E_A) - P(E_L, E_A)$, we can also write

$$UFAR = P(E_A) - P(E_L, E_A). \qquad (3\text{-}19)$$

---

[11] *UFAR* is often called *consumer's risk* in the statistics literature [17] - [19]. It has also been called the *probability of a false accept* or *PFA*.

3.2.2 Conditional False Accept Risk

CFAR is defined as the probability that an accepted attribute will be out-of-tolerance, i.e., that the event $\bar{E}_L$ will occur given that the event E_A has occurred:

$$CFAR = P(\bar{E}_L \mid E_A). \tag{3-20}$$

Using Eq. (3-10), this can be expressed as

$$CFAR = \frac{P(\bar{E}_L, E_A)}{P(E_A)}. \tag{3-21}$$

Then, using Eq. (3-16), we have

$$CFAR = \frac{P(E_A) - P(E_L, E_A)}{P(E_A)}$$

$$= 1 - \frac{P(E_L, E_A)}{P(E_A)}. \tag{3-22}$$

3.2.3 *UFAR* and *CFAR*

Since $P(E_A) < 1$, *CFAR* is always larger than *UFAR* for a given test or calibration. This can be seen by combining Eqs. (3-18) and (3-21) to get

$$CFAR = \frac{UFAR}{P(E_A)}. \tag{3-23}$$

It is interesting to note that when false accept risk (or consumer's risk) is discussed in journal articles, conference papers and in some risk analysis software, it is *UFAR* that is being referenced. This may seem odd in that *UFAR* is not, by definition, referenced to the consumer's perspective. The main reason for the use of *UFAR* as the "default" false accept risk is that the development of the concept of *CFAR* did not occur until the late 1970s [4]. By then, professionals working in the field had become practiced in thinking in terms of *UFAR* exclusively and reluctant to accept an alternative.

But, before we embrace the exclusive use of *CFAR* in defining false accept risk, it should be said that its use assumes that rejected UUT attributes are not adjusted or otherwise corrected. If rejected attributes are restored in some way and subsequently returned to service, a more involved definition of *CFAR* is needed. This definition is found in discussions in which *CFAR* can be equated with the probability that UUT attributes will be out-of-tolerance following testing or calibration. This probability is computed using what is termed the "post-test distribution" [13, 36, 37].

3.3 False Reject Risk

Another measure of the quality of calibration or testing is the probability that in-tolerance attributes will be rejected. This probability is called **false reject risk** (*FRR*) or **producer's risk**.[12]

---

[12] FRR has also been called the probability of a false reject or PFR.

Given the definitions of E_L and E_A, FRR is given by

$$FRR = P(E_L, \bar{E}_A). \tag{3-24}$$

Using Eq. (3-17), this can be written

$$FRR = P(E_L) - P(E_L, E_A). \tag{3-25}$$

3.4   Risk Analysis Alternatives

3.4.1   Process-Level Analysis [13]

With this alternative, risks are evaluated for each UUT attribute test point prior to testing or calibration by applying expected levels of UUT attribute in-tolerance probability and assumed calibration or test measurement process uncertainties. With process-level risk control, test limits called "guardband limits," if needed, are developed in advance and may be incorporated in calibration or test procedures. Measured values observed outside guardband limits may trigger some corrective action, such as adjustment or repair, or may be rejected and reduced in status or disposed.

3.4.2   Bench-Level Analysis

In addition to process-level risk control, bench-level methods are available to control risks in response to equipment attribute values obtained during test or calibration. With bench-level methods, guardband limits are superfluous, since corrective actions are triggered by on-the-spot risk or other measurement quality metric computation.

There are two bench-level alternatives: Bayesian analysis and confidence level analysis.

3.4.2.1   Bayesian Analysis

The Bayesian risk analysis methodology was developed by Castrup [28] and Jackson [29] in the '80s and later published with the label SMPC (Statistical Measurement Process Control) [30]. These methods enabled the analysis of false accept risk for UUT attributes, the estimation of both UUT attribute and MTE reference attribute biases, and the uncertainties in these biases.

Bayes' Theorem

The probability relations discussed previously lead us to an important expression referred to as Bayes' theorem. This theorem is of considerable value in computing measurement decision risks in test and calibration. Its derivation is simple and straightforward:

Returning to Eq. (3-11), we can write

$$P(E_A \mid E_L)P(E_L) = P(E_L \mid E_A)P(E_A), \tag{3-26}$$

which, after rearranging becomes

$$P(E_L \mid E_A) = \frac{P(E_A \mid E_L)P(E_L)}{P(E_A)}. \tag{3-27}$$

---

[13] Also referred to as "program-level" analysis [22].

Eq. (3-27) is Bayes' theorem in its simplest form.

In applying Bayes' theorem, a risk analysis will be performed for accepting a specific attribute based on *a priori* (pre-test) knowledge and on the results of measuring the attribute during testing or calibration.

The latter results comprise what is called "post-test" or *a posteriori* knowledge which, when combined with *a priori* knowledge, allow us to compute the quantities of interest, such as UUT and MTE attribute biases, bias uncertainties and pre-test in-tolerance probabilities. Obtaining these estimates is covered in Chapter 4 and Appendix C. The derivation of the expressions used in Bayesian analysis is given in Appendix D.

3.4.2.2 Confidence Level Analysis

Another bench-level approach, referred to as "confidence level analysis," evaluates the confidence that a measured UUT attribute value is in-tolerance, based on the uncertainty in the measurement process. Confidence level analysis is applied when an estimate of the *a priori* UUT attribute in-tolerance probability is not available. As such, it is not a true "risk control" method, but rather an application of the results of measurement uncertainty analysis. Confidence level analysis is discussed in detail in Sections 4.4 and 5.5.

3.5 The 4:1 TUR Alternative

Over the past few decades, the control of measurement decision risk has been embodied in requirements specifying the relative accuracy of the test or calibration process to the specified accuracy of the UUT attribute being tested or calibrated [31, 32]. These requirements provided some loose control of measurement decision risk but were not unambiguously defined or standardized. At the date of publication of this document, an explicit and rigorous relative accuracy requirement has been defined in ANSI/NCSL Z540.3-2006 [33].

Where it is not practical to compute false accept risk, the standard requires that the measurement's "test uncertainty ratio" or TUR, shall be greater than or equal to 4:1. The efficacy of this fallback is a matter of some contention, as is discussed in Section 3.5.2.

3.5.1 The Z540.3 Definition

Z540.3 defines TUR as the ratio of the span of the UUT tolerance to twice the "95%" expanded uncertainty of the measurement process used for calibration.[14] A caveat is provided in the form of a note stating that this requirement applies only to two-sided tolerances.

Mathematically, the 4:1 Z540.3 TUR definition is stated for tolerance limits $-L_1$ and L_2 as

$$\text{TUR} = \frac{L_1 + L_2}{2U_{95}}, \qquad (3\text{-}28)$$

where U_{95} is equal to the standard uncertainty u of the measurement process multiplied by a coverage factor k_{95} that corresponds to 95% confidence

---

14 See Appendix A for the definition of "expanded uncertainty."

$$U_{95} = k_{95}u. \tag{3-29}$$

In Z540.3, $k_{95} = 2$.

In addition to restricting the applicability of Eq. (3-28) to the calibration of UUT attributes with two-sided tolerance limits, Z540.3 also advises that Eq. (3-28) is strictly valid only in cases where the tolerance limits are symmetric, i.e., where $L_1 = L_2$. In such cases, the UUT attribute tolerance limits would be expressed in the form $\pm L$, and we would have

$$\text{TUR} = \frac{L}{U_{95}}. \tag{3-30}$$

3.5.2   A Critique of the 4:1 Requirement

When nominal uncertainty ratio requirements were originally developed [3], the computing machinery available at the time did not enable expedient cost-effective computation of measurement decision risk. Consequently, simple criteria were implemented that provided some measure of control.

In the present day, sufficient computing power is readily available and risk estimation methods are so well documented that it is difficult to understand why such a risk control criterion is still being implemented when risks can now be so easily computed and evaluated.

With this in mind, certain characteristics of the Z540.3 TUR requirement deserve mention.

- The requirement is merely a ratio of UUT tolerance limits relative to the expanded uncertainty of the measurement process. It is, at best, a crude risk control tool, i.e., one that does not control risks to any specified level. Moreover, in some cases, it may be superfluous. For instance, what if all UUT attributes of a given manufacturer/model are in-tolerance prior to test or calibration? In this case, the false accept risk is zero regardless of the TUR.

- The requirement is not applicable to all measurement scenarios. It does not apply when tolerances are asymmetric or single-sided.

- The requirement treats the expanded uncertainty $2u$ as a 95% confidence limit. This practice is not necessarily valid.[15] Appropriate methods of determining confidence limits are given in Appendix A and in References [26] and [27].

3.6   Recommendations

In many if not most cases, the single variable with the greatest impact on measurement decision risk is the *a priori* in-tolerance probability of the UUT attribute.[16] Consequently, definitions that fail to take this variable into account are *ipso facto* deficient. With this in mind, neither confidence level analysis nor the use of the nominal 4:1 TUR criterion are recommended unless absolutely nothing can be said about the UUT attribute's in-tolerance probability. If this is the case, confidence level analysis is recommended over the use of the 4:1 TUR criterion. While

---

[15] If the degrees of freedom for u is 15, the expanded uncertainty is roughly $2.13u$. As the degrees of freedom increases, the expanded uncertainty $2u$ becomes a better approximation of a 95% confidence limit.

[16] See Section 5.1 of Annex 4.

both the 4:1 criterion and confidence level analysis are easy to apply, the latter at least quantifies the degree to which a UUT attribute is in conformance with tolerances. The 4:1 criterion provides no such information. It serves to control false accept risk to some amount, but this amount is unknown and, possibly, insufficient.

If UUT attribute *a priori* in-tolerance information is available, either classical program-level analysis or Bayesian bench-level analysis is recommended. Of these, Bayesian analysis is preferable on the grounds that it provides a more explicit measure of false accept risk and offers on-the-spot information for deciding whether to adjust or otherwise correct a tested or calibrated attribute. In addition, if the probabilities on the right-hand side of Eq. (3-27) represent normally distributed quantities; and, if the statistics of the distributions can be estimated, the *a posteriori* in-tolerance probability of the UUT attribute can be computed with commercial spreadsheet applications without the need for any additional programming.

Chapter 4: Computing Risk

In this chapter, methods for computing measurement decision risk are presented within the framework of both process-level and bench-level analyses. Process-level analysis employs what is referred to as "the classical method." Bench-level analyses include methods referred to as the "Bayesian method" and the "confidence level method." The classical method and the Bayesian method employ the probability relations developed in Chapter 3. The confidence level method employs calibration uncertainties and conventional statistics.

Each method computes risks using the results of UUT calibrations, which are performed within the context of four calibration scenarios. Calibration results are employed to estimate unconditional false accept risk (*UFAR*), conditional false accept risk (*CFAR*) and the false reject risk (*FRR*), as defined in Chapter 3. Examples are given in Appendix A to illustrate concepts and procedures.

4.1   Calibration Scenarios

The four calibration scenarios are:

1. The measurement reference (MTE) measures the value of an attribute of the unit under test (UUT) that provides an output or stimulus.

2. The UUT measures the value of a reference attribute of the MTE that provides an output or stimulus.

3. The UUT and MTE each provide an output or stimulus for comparison using a bias cancellation comparator.

4. The UUT and MTE both measure the value of an attribute of a common device or artifact that provides an output or stimulus.

The results of calibration include an observed value, referred to as a "measurement result" or "calibration result," and an estimated calibration uncertainty. For each scenario, a measurement equation is given that is applicable to the manner in which calibrations are performed and calibration results are recorded or interpreted.

The detailed development of each scenario, identification of measurement errors, and the computation of uncertainties are given in Appendix A, along with discussions of related concepts and definitions of terms.

4.1.1   Risk Variables

For each scenario, the basic set of variables that are important for measurement decision risk analysis are shown in Table 4-1.

Table 4-1. Risk Variables Nomenclature.

Variable	Definition
UUT	the unit under test or calibration
MTE	the measurement reference standard used to calibrate the UUT
x	(1) the value of the UUT attribute at the time of calibration or (2) a measured value of the MTE reference attribute obtained by the UUT attribute
x_n	the nominal value of the UUT attribute
y	(1) the value of the MTE reference attribute at the time of calibration or a (2) measured value of the UUT attribute obtained by the MTE reference attribute
y_n	the nominal value of the MTE reference attribute
$e_{UUT,b}$	the bias of the UUT attribute value at the time of calibration
$u_{UUT,b}$	the uncertainty in $e_{UUT,b}$, i.e., the standard deviation of the probability distribution of the population of $e_{UUT,b}$ values.[17]
$-L_1$ and L_2	the tolerance limits for $e_{UUT,b}$
$-A_1$ and A_2	the "acceptance" limits (test limits) for $e_{UUT,b}$
L	the range of values of $e_{UUT,b}$ from $-L_1$ to L_2 (the UUT tolerance limits)
$\mathcal{A}$	the range of values of $e_{UUT,b}$ from $-A_1$ to A_2 (the UUT acceptance limits)
δ	a measurement (estimate) of $e_{UUT,b}$ obtained through calibration
ε_{cal}	the total error in δ
u_{cal}	the uncertainty in ε_{cal}, i.e., the uncertainty in the value of δ

These variables will be employed in various probability relations in the next section.[18]

4.1.2   Probability Relations

The fundamental probability functions of measurement decision risk analysis are constructed in this section. In constructing these functions, we make use of the notation of mathematical logic and set theory, in which the $\in$ operator reads "belongs to" or "is "included in." Likewise, the $\notin$ operator reads "does not belong to" or "is excluded from." The notation of logic and set theory notation is briefly discussed in Appendix I. Using the nomenclature in Table 4-1, the probability functions are given in Table 4-2.

---

[17] See Appendix A, reference [1] or reference [2].

[18] Cases where the UUT attribute has a single-sided upper or lower tolerance limit are accommodated by setting one of the tolerance limits in Table 4-1 to an applicable limiting physical value. For example, for a single-sided upper limit, with an essentially unbounded lower limit, L_1 and A_1 would be set to ∞. For a single-sided lower limit, with an essentially unbounded upper limit, L_2 and A_2 would be set to ∞.

Table 4-2. Risk Computation Nomenclature.

Risk Analysis Function	Definition
$P(e_{UUT,b} \in \mathsf{L})$	the *a priori* probability that $-L_1 \leq e_{UUT,b} \leq L_2$. This is the probability that the UUT attribute to be calibrated is in-tolerance at the time of calibration.
$P(\delta \in \mathcal{A})$	the probability that $-A_1 \leq \delta \leq A_2$. This is the probability that measured values of $e_{UUT,b}$ will be accepted as being in-tolerance.
$P(e_{UUT,b} \in \mathsf{L}, \delta \in \mathcal{A})$	the probability that $-L_1 \leq e_{UUT,b} \leq L_2$ and $-A_1 \leq \delta \leq A_2$. This is the joint probability that a UUT attribute will be in-tolerance *and* will be observed to be in-tolerance.
$P(\delta \in \mathcal{A} \mid e_{UUT,b} \in \mathsf{L})$	the probability that, if $-L_1 \leq e_{UUT,b} \leq L_2$, then $-A_1 \leq \delta \leq A_2$. This is the conditional probability that an in-tolerance attribute will be accepted as in-tolerance.
$P(e_{UUT,b} \notin \mathsf{L}, \delta \in \mathcal{A})$	the probability that $e_{UUT,b}$ lies outside $\mathcal{L}$ and $-A_1 \leq \delta \leq A_2$. This is the joint probability that a UUT attribute will be out-of-tolerance *and* will be observed to be in-tolerance.
$P(e_{UUT,b} \in \mathsf{L}, \delta \notin \mathcal{A})$	the probability that $-L_1 \leq e_{UUT,b} \leq L_2$ and δ lies outside $\mathcal{A}$. This is the joint probability that a UUT attribute will be in-tolerance *and* will be observed to be out-of-tolerance.
$P(e_{UUT,b} \notin \mathsf{L} \mid \delta \in \mathcal{A})$	the probability that $e_{UUT,b}$ lies outside $\mathcal{L}$ *given* that $-A_1 \leq \delta \leq A_2$. The conditional probability that an accepted UUT attribute will be out-of-tolerance.

Table 4-3 shows the equivalence of the probability functions in Table 4-2 with the probability functions of Chapter 3. In Table 4-3, the variable E_L represents the event that the UUT attribute is in-tolerance and E_A represents the event that the attribute is observed to be in-tolerance.

Table 4-3. Correspondence between the Risk Analysis Nomenclature and the Probability Functions of Chapter 3.

Risk Analysis Function	Basic Probability Representation
$P(e_{UUT,b} \in \mathsf{L})$	$P(E_L)$
$P(\delta \in \mathcal{A})$	$P(E_A)$
$P(e_{UUT,b} \in \mathsf{L}, \delta \in \mathcal{A})$	$P(E_L, E_A)$
$P(\delta \in \mathcal{A} \mid e_{UUT,b} \in \mathsf{L})$	$P(E_A \mid E_L)$
$P(e_{UUT,b} \notin \mathsf{L}, \delta \in \mathcal{A})$	$P(\bar{E}_L, E_A)$
$P(e_{UUT,b} \in \mathsf{L}, \delta \notin \mathcal{A})$	$P(E_L, \bar{E}_A)$
$P(e_{UUT,b} \notin \mathsf{L} \mid \delta \in \mathcal{A})$	$P(\bar{E}_L \mid E_A)$

4.1.3 Calibration Scenario Results

Using the nomenclature in Table 4-3, the measurement result δ is defined for each scenario as shown in Table 4-4.

Table 4-4. Measurement Result Definitions for the Calibration Scenarios.

Scenario	Description	δ	Comment
1	MTE measures UUT	$y - x_n$	y is the measured value of a UUT attribute; x_n is the attribute's nominal value
2	UUT measures MTE	$x - y_n$	x is the measured value of an MTE attribute; y_n is the attribute's reference or nominal value
3	UUT and MTE attribute values are obtained using a comparator	$x - y$	x and y are UUT and MTE measurements
4	UUT and MTE measure a common attribute	$x - y$	x and y are UUT and MTE measurements

As stated earlier, the calibration error for each scenario is denoted ε_{cal} and the uncertainty in this error is u_{cal}. The specific combinations of measurement process errors comprising ε_{cal} are described in Appendix A. The uncertainty u_{cal} is obtained by taking the statistical variance of ε_{cal}:

$$u_{cal} = \sqrt{\text{var}(\varepsilon_{cal})} .$$

4.2   The Classical Method

The classical method provides process level decision risk control in that risk estimation can assist in making equipment adjustment or repair decisions using nominal criteria, such as guardbands. Estimates obtained using the classical method are also useful in making equipment procurement decisions, adjusting calibration intervals, and setting end-of-period measurement reliability targets.[19]

The fundamentals of the classical method are given in the following sections. Detailed discussion and derivations are given in Appendix B.

4.2.1   Measurement Decision Risk Recap

To reiterate from Chapter 3, the risk definitions employed in the classical method are

$$UFAR = P(\bar{E}_L, E_A)$$
$$= P(E_A) - P(E_L, E_A), \tag{4-1}$$

$$CFAR = P(\bar{E}_L \mid E_A)$$
$$= 1 - \frac{P(E_L, E_A)}{P(E_A)}, \tag{4-2}$$

and

$$FRR = P(E_L, \bar{E}_A)$$
$$= P(E_L) - P(E_L, E_A). \tag{4-3}$$

---

[19] Setting end-of-period reliability targets is discussed in Appendix H.

Expressions for computing the probability functions of these definitions are given in the next section.

4.2.2   Estimating Risk

4.2.2.1   Relevant Functions

The probability functions in Table 4-3 can be constructed using the probability distributions described in Appendix A. These distributions are mathematically represented by probability density functions (pdfs) that relate random variables of interest to their probability of occurrence. Table 4-5 defines the relevant pdfs.

Table 4-5. Risk Analysis Probability Density Functions[20]

pdf	Description
$f(e_{UUT,b})$	pdf for the UUT bias at the time of calibration
$f(\delta)$	pdf for the measurement result
$f(\delta, e_{UUT,b})$	pdf for the joint distribution of δ and $e_{UUT,b}$
$f(\delta \mid e_{UUT,b})$	pdf for the conditional distribution of δ given a value of $e_{UUT,b}$
$f(e_{UUT,b} \mid \delta)$	pdf for the conditional distribution of $e_{UUT,b}$ given a value of δ

Using the cross-references of Tables 4-2 through 4-4, the basic probability functions used in the classical method can be written

$$P(E_L) = \int_{-L_1}^{L_2} f(e_{UUT,b}) \, de_{UUT,b} \,, \tag{4-4}$$

$$P(E_L, E_A) = \int_{-L_1}^{L_2} \int_{-A_1}^{A_2} f(\delta, e_{UUT,b}) \, d\delta \, de_{UUT,b}$$

$$= \int_{-L_1}^{L_2} \int_{-A_1}^{A_2} f(\delta \mid e_{UUT,b}) f(e_{UUT,b}) \, d\delta \, de_{UUT,b} \,, \tag{4-5}$$

and

$$P(E_A) = \int_{-\infty}^{\infty} \int_{-A_1}^{A_2} f(\delta, e_{UUT,b}) \, d\delta \, de_{UUT,b}$$

$$= \int_{-\infty}^{\infty} \int_{-A_1}^{A_2} f(\delta \mid e_{UUT,b}) f(e_{UUT,b}) \, d\delta \, de_{UUT,b} \,, \tag{4-6}$$

where use was made of Eq. (3-10).

---

[20] To be more rigorous with respect to notation, each pdf would have its own letter designator or subscript to distinguish its functional form from other pdfs. Such rigor is laudable but leads to a more tedious notation than we already have. It is hoped that the distinct character of each pdf will be apparent from its context of usage.

4.2.2.2   Risk Estimation

In classical risk analysis, it is ordinarily assumed that the measurement result δ is normally distributed with a mean value of $e_{UUT,b}$ and a standard deviation of u_{cal}. Then we can write

$$
\begin{aligned}
P(E_L, E_A) &= \frac{1}{\sqrt{2\pi}u_{cal}} \int_{-L_1}^{L_2} \int_{-A_1}^{A_2} e^{-(\delta - e_{UUT,b})^2/2u_{cal}^2} f(e_{UUT,b}) \, d\delta \, de_{UUT,b} \\
&= \int_{-L_1}^{L_2} \left[\Phi\left(\frac{A_2 - e_{UUT,b}}{u_{cal}} \right) + \Phi\left(\frac{A_1 + e_{UUT,b}}{u_{cal}} \right) - 1 \right] f(e_{UUT,b}) \, de_{UUT,b},
\end{aligned}
\tag{4-7}
$$

and

$$
\begin{aligned}
P(E_A) &= \frac{1}{\sqrt{2\pi}u_{cal}} \int_{-\infty}^{\infty} \int_{-A_1}^{A_2} f(e_{UUT,b}) e^{-(\delta - e_{UUT,b})^2/2u_{cal}^2} \, d\delta \, de_{UUT,b} \\
&= \int_{-\infty}^{\infty} \left[\Phi\left(\frac{A_2 - e_{UUT,b}}{u_{cal}} \right) + \Phi\left(\frac{A_1 + e_{UUT,b}}{u_{cal}} \right) - 1 \right] f(e_{UUT,b}) \, de_{UUT,b},
\end{aligned}
\tag{4-8}
$$

where Φ is the normal distribution function available in most spreadsheet applications.

The variable $e_{UUT,b}$ may follow any number of plausible probability distributions. A sample is given in Appendix A. In all cases, $e_{UUT,b}$ is assumed to have a zero mean value and a standard deviation of $u_{UUT,b}$. Like the variable δ, $e_{UUT,b}$ is often assumed to be normally distributed. For such cases, we have

$$
\begin{aligned}
P(E_L) &= \frac{1}{\sqrt{2\pi}u_{UUT,b}} \int_{-L_1}^{L_2} e^{-e_{UUT,b}^2/2u_{UUT,b}^2} de_{UUT,b} \\
&= \Phi\left(\frac{L_1}{u_{UUT,b}} \right) + \Phi\left(\frac{L_2}{u_{UUT,b}} \right) - 1,
\end{aligned}
\tag{4-9}
$$

$$
P(E_L, E_A) = \frac{1}{\sqrt{2\pi}u_{UUT,b}} \int_{-L_1}^{L_2} \left[\Phi\left(\frac{A_2 - e_{UUT,b}}{u_{cal}} \right) + \Phi\left(\frac{A_1 + e_{UUT,b}}{u_{cal}} \right) - 1 \right] e^{-e_{UUT,b}^2/2u_{UUT,b}^2} \, de_{UUT,b},
\tag{4-10}
$$

and

$$
\begin{aligned}
P(E_A) &= \frac{1}{2\pi u_{UUT,b} u_{cal}} \int_{-\infty}^{\infty} \int_{-A_1}^{A_2} e^{-(\delta - e_{UUT,b})^2/2u_{cal}^2} e^{-e_{UUT,b}^2/2u_{UUT,b}^2} \, d\delta \, de_{UUT,b} \\
&= \frac{1}{\sqrt{2\pi}u_A} \int_{-A}^{A_2} e^{-\delta^2/2u_A^2} \, d\delta \\
&= \Phi\left(\frac{A_1}{u_A} \right) + \Phi\left(\frac{A_2}{u_A} \right) - 1,
\end{aligned}
\tag{4-11}
$$

where

$$
u_A = \sqrt{u_{UUT,b}^2 + u_{cal}^2}.
\tag{4-12}
$$

4.2.2.3 Obtaining the Required Parameters

Obtaining the variable δ and estimating the uncertainty u_{cal} are described in Appendix A. The standard deviation $u_{UUT,b}$ may comprise a Type A estimate but is usually obtained as a Type B estimate in which the containment limits are $-L_1$ and L_2 and the containment probability is taken to be the percent of UUT attributes observed to be in-tolerance at the time of the UUT test or calibration. Containment limits and containment probability are defined in Appendix A. For a more complete discussion of these quantities and of Type B analysis in general, see Annex 3 or References [26] or [27].

To summarize, the information needed for computing the probability functions used in classical measurement decision risk analysis consists of the following:

- The limits $-L_1$ and L_2
- The containment probability for $e_{UUT,b}$ at the end of the UUT calibration interval
- The distribution for $e_{UUT,b}$ (see Appendix A)
- The limits $-A_1$ and A_2
- The calibration result δ
- The calibration uncertainty u_{cal}

4.2.2.4 UUT In-Tolerance Probability

In the foregoing, the in-tolerance probability (containment probability) of the UUT attribute is used in estimating the quantity $u_{UUT,b}$. Ideally, the in-tolerance probability would be obtained at the test point (attribute) level, i.e., the point at which the UUT attribute is calibrated. However, for many testing or calibrating organizations, in-tolerance probability information is available only as a percent in-tolerance at the UUT item (serial number) level or higher. The relationship between item level or higher and test-point level is shown in Figure 4-1.

Ordinarily, a UUT item is declared in-tolerance only if all test points are found to be in-tolerance. Since the probability of the joint occurrence of two or more events is lower than the occurrence of any individual constituent event, see Section 3.1, it can be seen that the in-tolerance probability at each test point must be inherently greater than the reported in-tolerance probability for the item. Then computing $u_{UUT,b}$ using the reported percent in-tolerance at time of calibration will yield a value that is larger than what is appropriate at the test point level. This "inflated" value results in measurement decision risk estimates that are likewise inflated. Consequently, if such estimated risks are acceptable, it follows that whatever risks are present at the test point level are also acceptable.

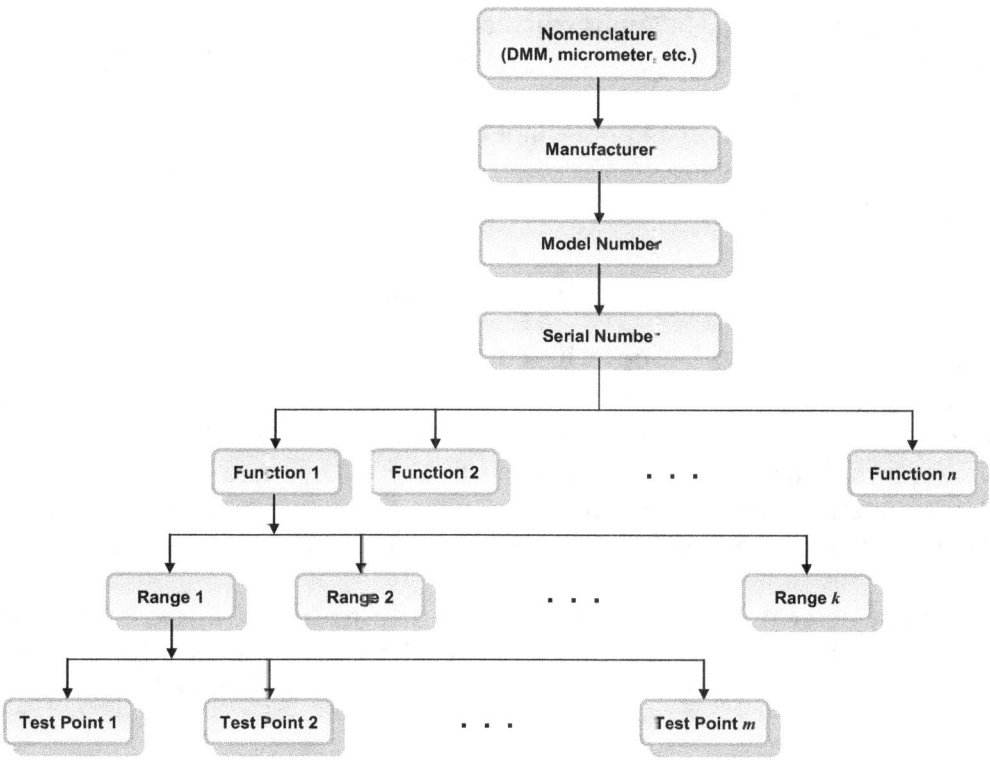

Figure 4-1. UUT/MTE Hierarchy

Shown is a case of a manufacturer/model with n functions. Function 1 has k ranges and range 1 is calibrated at m test points. The term "parameter" is sometimes used in place of "function."

In some cases, it may be feasible to estimate the test point in-tolerance probabilities in terms of the item level in-tolerance probability. Specifically, if an item is calibrated at q independent test points, each with an inherently equal in-tolerance probability r, then the item level in-tolerance probability R can be expressed as

$$R = r^q,$$

from which

$$r = R^{1/q}.$$

As an example, consider a gage block set of 10 gage blocks. The results of calibration for such an item are typically recorded for the set as a whole. Accordingly, the set is declared in-tolerance only if all 10 gage blocks are in-tolerance. Imagine that the reported percent in-tolerance for the set yields an estimated in-tolerance probability of $R = 0.92$. If it can be assumed, by virtue of similar fabrication and materials that each gage block is characterized by the same in-tolerance probability r, then the in-tolerance probability at the test point level is computed to be

$$r \cong R^{1/10}$$
$$= (0.92)^{1/10} = 0.991,$$

considerably higher than 0.92. The "approximately equals" sign indicates that other factors, such as usage rate, inhomogeneity with respect to tolerance limits, etc. may be in play.

The difference between item level and test-point level in-tolerance probability increases with increasing values of q. For instance, if the gage block set describe above is comprised of 30 individual gage blocks, then

$$r \cong R^{1/30}$$
$$= (0.92)^{1/30} = 0.997.$$

4.2.2.5   True vs. Reported Percent In-Tolerance

There is an additional wrinkle in estimating UUT attribute in-tolerance probability. Because of false accept and false reject risks, the observed (reported) percent in-tolerance will be consistently lower than the true percent in-tolerance. For cases where $e_{UUT,b}$ is not normally distributed, calculating the true percent in-tolerance from the reported value is somewhat difficult. Both normal and non-normal cases are discussed in detail in Appendix E.

Calculating the true percent in-tolerance is simple for a normally distributed $e_{UUT,b}$ with symmetric tolerance limits $\pm L$. We do this by taking advantage of the fact that the reported percent in-tolerance R is equated with the probability $P(E_A)$. From Eq. (4-11), this probability is given by

$$P(E_A) = \frac{1}{2\pi u_{UUT,b} u_{cal}} \int_{-\infty}^{\infty} \int_{-L}^{L} e^{-(\delta - e_{UUT,b})^2 / 2u_{cal}^2} e^{-e_{UUT,b}^2 / 2u_{UUT,b}^2} \, d\delta \, de_{UUT,b}$$

$$= \frac{1}{\sqrt{2\pi} u_A} \int_{-L}^{L} e^{-\delta^2 / 2u_A^2} \, d\delta \qquad (4\text{-}13)$$

$$= 2\Phi\left(\frac{L}{u_A}\right) - 1,$$

where the adjustment limits A_1 and A_2 have been set equal to the tolerance limits, and

$$u_A = \sqrt{u_{UUT,b}^2 + u_{cal}^2} \, . \qquad (4\text{-}14)$$

The variable u_A can be solved for from Eq. (4-13).

$$u_A = \frac{L}{\varphi}, \qquad (4\text{-}15)$$

where

$$\varphi = \Phi^{-1}\left[\frac{1 + P(E_A)}{2}\right], \qquad (4\text{-}16)$$

and Φ^{-1} is the inverse normal distribution function. Setting Eq. (4-14) equal to (4-15) gives the estimated true value of the UUT population standard deviation

$$u_{UUT,b} = \sqrt{u_A^2 - u_{cal}^2}$$
$$= \sqrt{(L / \varphi)^2 - u_{cal}^2} \, . \qquad (4\text{-}17)$$

4.3   The Bayesian Method

As stated earlier, with the Bayesian method, false accept and false reject risks are estimated using the measured value or sample mean obtained from a given test or calibration. Using this method, the testing or calibration result can be entered in a spreadsheet or software program and the relevant decision risks are calculated on the spot. For this reason, the Bayesian method is classified as a bench-level method.

The fundamentals of the Bayesian method are presented in the following sections. An introduction to Bayesian concepts is provided in Appendix C. Derivations of the expressions used in this Handbook Annex are given in Appendix D.

> **Note**: The Bayesian method described herein is applicable when attribute biases are normally distributed.

4.3.1   Risk Analysis for a Measured Variable

The procedure for applying Bayesian methods to perform risk analysis for a measured attribute is as follows:

1. Assemble all relevant *a priori* knowledge, such as the tolerance limits for the UUT attribute, the tolerance limits for the reference attribute, the in-tolerance probabilities for each attribute and the uncertainty of the measuring process.

2. Perform a measurement or set of measurements. This may take place with any of the four calibration scenarios of Appendix A.

3. Estimate the UUT attribute and reference attribute biases using Bayesian methods.

4. Compute uncertainties in the bias estimates.

5. Act on the results. Report the biases and bias uncertainties, along with in-tolerance probabilities for the attributes, and adjust each attribute to correct the estimated biases, as appropriate.

4.3.2   *A priori* Knowledge

The *a priori* knowledge for a Bayesian analysis may include several kinds of information. For example, if the UUT attribute is the pressure of an automobile tire, such knowledge may include a rigorous projection of the degradation of the tire's pressure as a function of time since the tire was last inflated or a crude estimate based on the appearance of the tire's lateral bulge. However *a priori* knowledge is obtained, it should lead to the following quantities:

- Estimates of the uncertainties in the biases of both the UUT attribute and the reference attribute. These estimates may be obtained by Type B analysis using containment limits and containment probabilities or by other means, if applicable.

- An estimate of the uncertainty due to measurement process error other than the bias in the reference attribute, accounting for all error sources.

4.3.3 Post-Test Knowledge

The post-test knowledge in a Bayesian analysis consists of the results of measurement. As stated earlier, these results may be in the form of a measurement or a set of measurements. The measurements may be measurements of the UUT attribute value in the form of readings provided by the reference attribute, readings provided by the UUT attribute from measurements of the reference attribute, or readings provided by both the UUT attribute and reference attribute, taken on a common device or artifact.

4.3.4 Bias Estimates

UUT attribute and reference attribute biases are estimated using the method described in Appendix D. The method encompasses cases where a measurement sample is taken by the UUT attribute, the reference attribute or both.[21] The variables are given in Table 4-6.

Table 4-6. Bayesian Estimation Variables.

Variable	Description
$e_{UUT,b}$	UUT attribute bias at the time of calibration
$u_{UUT,b}$	UUT attribute bias standard uncertainty
δ	UUT attribute calibration result, as defined in Appendix A
$e_{MTE,b}$	MTE reference attribute bias at the time of calibration
$u_{MTE,b}$	MTE attribute bias standard uncertainty
u_{cal}	uncertainty in the UUT attribute calibration process, as defined in Appendix A
$-L_1$ and L_2	lower and upper UUT attribute tolerance limits
$-l_1$ and l_2	lower and upper MTE reference attribute tolerance limits

4.3.4.1 UUT Bias

Employing the pdfs of Table 4-5 in Bayes' theorem, given in Eq. (3-27), gives Bayes' relation for the pdf of interest

$$f(e_{UUT,b} \mid \delta) = \frac{f(\delta \mid e_{UUT,b}) f(e_{UUT,b})}{f(\delta)}, \qquad (4\text{-}18)$$

where

$$f(\delta) = \int_{-\infty}^{\infty} f(\delta, e_{UUT,b}) \, de_{UUT,b}$$

$$= \int_{-\infty}^{\infty} f(\delta \mid e_{UUT,b}) f(e_{UUT,b}) \, de_{UUT,b}. \qquad (4\text{-}19)$$

For normally distributed δ and $e_{UUT,b}$, Eq. (4-19) becomes

---

[21] Actually, the methodology described in Appendix D can also be applied to measurements of a quantity made by any number of devices.

$$f(\delta) = \int_{-\infty}^{\infty} f(\delta, e_{UUT,b}) \, de_{UUT,b}$$

$$= \frac{1}{2\pi u_{cal} u_{UUT,b}} \int_{-\infty}^{\infty} e^{-(\delta - e_{UUT,b})^2 / 2u_{cal}^2} e^{-e_{UUT,b}^2 / 2u_{UUT,b}^2} \, de_{UUT,b} \qquad (4\text{-}20)$$

$$= \frac{1}{\sqrt{2\pi} u_A} e^{-\delta^2 / 2u_A^2},$$

where, as before,

$$u_A = \sqrt{u_{UUT,b}^2 + u_{cal}^2} \,.$$

Substituting this pdf into Eq. (4-18), together with

$$f(\delta \mid e_{UUT,b}) = \frac{1}{\sqrt{2\pi} u_{cal}} e^{-(\delta - e_{UUT,b})^2 / 2u_{cal}^2} \qquad (4\text{-}21)$$

and

$$f(e_{UUT,b}) = \frac{1}{\sqrt{2\pi} u_{UUT,b}} e^{-e_{UUT,b}^2 / 2u_{UUT,b}^2}, \qquad (4\text{-}22)$$

yields

$$f(e_{UUT,b} \mid \delta) = \frac{1}{\sqrt{2\pi} u_\beta} \exp\left\{ -\left[(\delta - e_{UUT,b})^2 / 2u_{cal}^2 + e_{UUT,b}^2 / 2u_{UUT,b}^2 - \delta^2 / 2u_A^2 \right] \right\}$$

$$= \frac{1}{\sqrt{2\pi} u_\beta} e^{-(e_{UUT,b} - \beta)^2 / 2u_\beta^2}, \qquad (4\text{-}23)$$

where

$$\beta = \frac{u_{UUT,b}^2}{u_A^2} \delta \qquad (4\text{-}24)$$

and

$$u_\beta = \frac{u_{UUT,b} u_{cal}}{u_A}. \qquad (4\text{-}25)$$

Given these results, along with the properties of the normal distribution, we see that β is the estimated value for $e_{UUT,b}$ and u_β is the estimated bias uncertainty:

$$\text{UUT Attribute Bias} = \beta = \frac{u_{UUT,b}^2}{u_A^2} \delta, \qquad (4\text{-}26)$$

and

$$\text{UUT Attribute Bias Uncertainty} = u_\beta = \frac{u_{UUT,b}}{u_A} u_{cal}. \qquad (4\text{-}27)$$

4.3.4.2 MTE Bias

With the Bayesian method, calibration results can be used to obtain an estimate of the bias of the calibration reference attribute and the uncertainty in this estimate. This is accomplished by imagining that the UUT is calibrating the MTE. We begin by replacing $u_{UUT,b}$ with $u_{MTE,b}$ and δ with $-\delta$ in Eq. (4-26) and by defining a variable α

$$\text{MTE Attribute Bias} = \alpha = -\frac{u_{MTE,b}^2}{u_A^2}\delta\ . \tag{4-28}$$

The first step in estimating the uncertainty in this bias is to define a new uncertainty term

$$u_{process} = \sqrt{u_{cal}^2 - u_{MTE,b}^2}\ . \tag{4-29}$$

Next, a calibration uncertainty is defined that would apply if the UUT were calibrating the MTE:

$$u_{cal}' = \sqrt{u_{UUT,b}^2 + u_{process}^2}\ . \tag{4-30}$$

Using this quantity in Eq. (4-27) yields the bias uncertainty u_α of the reference attribute

$$\text{MTE Attribute Bias Uncertainty} = u_\alpha = \frac{u_{MTE,b}}{u_A}u_{cal}'\ . \tag{4-31}$$

4.3.5   UUT Attribute In-Tolerance Probability

An estimate of the UUT attribute in-tolerance probability $P_{UUT,in}$ is obtained by integrating $f(e_{UUT,b}\mid\delta)$ from $-L_1$ to L_2

$$
\begin{aligned}
P_{UUT,in} &= \frac{1}{\sqrt{2\pi}u_\beta}\int_{-L_1}^{L_2} e^{-(e_{UUT,b}-\beta)^2/2u_\beta^2}\,de_{UUT,b}\\
&= \Phi\!\left(\frac{L_1+\beta}{u_\beta}\right) + \Phi\!\left(\frac{L_2-\beta}{u_\beta}\right) - 1,
\end{aligned}
\tag{4-32}
$$

where β is given in Eq. (4-26).

4.3.6   MTE Attribute In-Tolerance Probability

Since we have the necessary expressions at hand, we can also estimate the in-tolerance probability $P_{MTE,in}$ of the MTE is obtained by integrating the pdf $f(e_{MTE,b}\mid\delta)$ for the MTE bias from $-l_1$ to l_2

$$
\begin{aligned}
P_{MTE,in} &= \frac{1}{\sqrt{2\pi}u_\alpha}\int_{-l_1}^{l_2} e^{-(e_{MTE,b}-\alpha)^2/2u_\alpha^2}\,de_{MTE,b}\\
&= \Phi\!\left(\frac{l_1+\alpha}{u_\alpha}\right) + \Phi\!\left(\frac{l_2-\alpha}{u_\alpha}\right) - 1,
\end{aligned}
$$

where α is given in Eq. (4-28).

4.3.7   Bayesian False Accept Risk
4.3.7.1   Uncorrected UUT attribute

If the UUT attribute is accepted without adjustment, the false accept risk is just

$$CFAR = 1 - P_{UUT,in}.$$ (4-33)

The Bayesian false accept risk is labeled $CFAR$ because the pdf used to compute P_{in} is a conditional pdf $f(e_{UUT,b} \mid \delta)$.

4.3.7.2 Corrected UUT attribute

If the UUT attribute bias is corrected by adjustment or other means, the false accept risk will be reduced. Essentially, the situation is equivalent to a case where $\delta = 0$. Then $P_{UUT,in}$ becomes

$$P_{UUT,in} = \Phi\left(\frac{L_1}{u_{\beta,adj}}\right) + \Phi\left(\frac{L_2}{u_{\beta,adj}}\right) - 1,$$ (4-34)

where $u_{\beta,adj}$ is u_β modified to include uncertainties due to errors arising from adjustment or other corrective action.

4.4 The Confidence Level Method

As with the Bayesian method, using the confidence level method involves entering the results of testing or calibration in a spreadsheet or other program and obtaining analysis results. Since the data are entered and results obtained by testing or calibration personnel, the confidence level method is also a bench-level method.

With the confidence level method of analysis, the confidence that a UUT attribute value lies within its tolerance limits is computed. The confidence level method is distinguished from the classical and Bayesian methods in that the result of the analysis is an in-tolerance confidence level, rather than an in-tolerance probability. The method is applied when an estimate of the UUT attribute in-tolerance probability is not feasible. As such, it lacks information needed to compute measurement decision risk and, therefore, is not a true "risk control" method, but rather a "pseudo risk control" method.

4.4.1 Confidence Level Estimation

Confidence level estimation employs the variables shown in Table 4-7.

Table 4-7. Confidence Level Estimation Variables.

Variable	Description
δ	the UUT attribute calibration result, **as** defined in Appendix A
ζ	a random variable representing values of the population from which δ was obtained
u_{cal}	the uncertainty in the UUT attribute calibration process, as defined in Appendix A
$-L_1$ and L_2	the lower and upper UUT attribute tolerance limits

With the confidence level method, we assume that the random variable ζ is normally distributed with mean δ and standard deviation u_{cal}. Then, given a calibration result δ, the in-tolerance confidence level is obtained from[22]

$$P_{UUT,in} = \frac{1}{\sqrt{2\pi}u_{cal}} \int_{-L_1}^{L_2} e^{-(\zeta-\delta)^2/2u_{cal}^2} d\zeta$$

$$= \Phi\left(\frac{L_1+\delta}{u_{cal}}\right) + \Phi\left(\frac{L_2-\delta}{u_{cal}}\right) - 1. \tag{4-35}$$

4.4.2 Applying Confidence Level Estimates

As with the Bayesian method, corrective action may be called for if a computed confidence level P_{in} is less than a predetermined specified limit. Let the maximum allowable risk be denoted r_{max}. Then corrective action is called for if $P_{UUT,in} < 1 - r_{max}$. If the Z540.3 requirement is adhered to, $r_{max} = 0.02$.

4.4.3 UUT Attribute Adjustment

Adjustment of the UUT attribute sets the value of the variable δ to zero. Hence Eq. (4-35) would be replaced by

$$P_{UUT,in} = \Phi\left(\frac{L_1}{u_{cal,adj}}\right) + \Phi\left(\frac{L_2}{u_{cal,adj}}\right) - 1, \tag{4-36}$$

where $u_{cal,adj}$ is u_{cal} modified to include uncertainties due to errors arising from adjustment or other corrective action.

---

[22] As with the Bayesian method, cases where the UUT attribute has a single-sided upper or a single-sided lower tolerance limit are accommodated by setting $L_1 = \infty$ or $L_2 = \infty$, respectively.

Chapter 5: Compensating Measures

When risks exceed allowable amounts, several steps may be taken to either alleviate them or compensate for them. The principal measures that are customarily turned to for such compensation are

- Increasing the in-tolerance probabilities of the UUT attribute
- Reducing the uncertainty of the measuring process
- Applying sequential acceptance testing
- Applying test guardbands

5.1   Increasing UUT In-Tolerance Probability

As Figure 5-1 shows, both false accept and false reject risks are sensitive to the in-tolerance probability of the UUT attribute.[23]

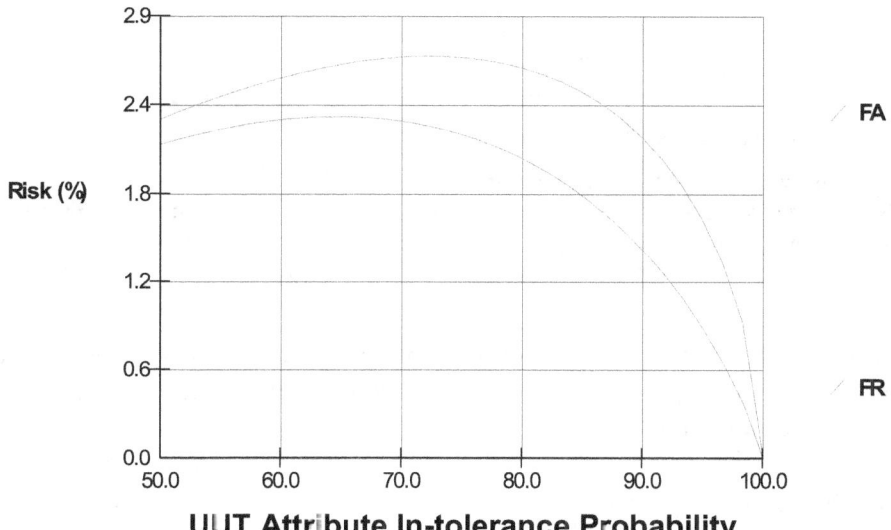

Figure 5-1. Risk vs. UUT Attribute In-Tolerance Probability.

Shown is a case where the measurement process standard uncertainty $u = 12.7553$, and the TUR = 4:1. As the figure illustrates, both false accept risk (FA) and false reject risk (FR) are functions of UUT attribute in-tolerance probability. Note that false accept risk decreases with increasing in-tolerance probability, while false reject risk exhibits the same behavior for higher in-tolerance probabilities but opposite behavior for lower in-tolerance probabilities. This is due to the fact that, for the latter, there are fewer in-tolerance attributes to falsely reject. In the plot, FA corresponds to *UFAR*.[24]

---

[23] The plots in Figure 5-1 and 5-2 were developed using AccuracyRatio 1 6 [43].

[24] Plots of *UFAR* show that it is not monotonic with respect to UUT attribute in-tolerance probability. It actually *decreases* below a pivotal probability whose value is dependent on the details of the test or calibration. This is because, below the pivotal value, UUT attribute biases that lie well outside the tolerance limits become more easily rejected by the test or calibration process.

5.2 Reducing Measurement Uncertainty

Figure 5-2 provides an example showing the relationship between measurement process uncertainty and both false accept and false reject risks.

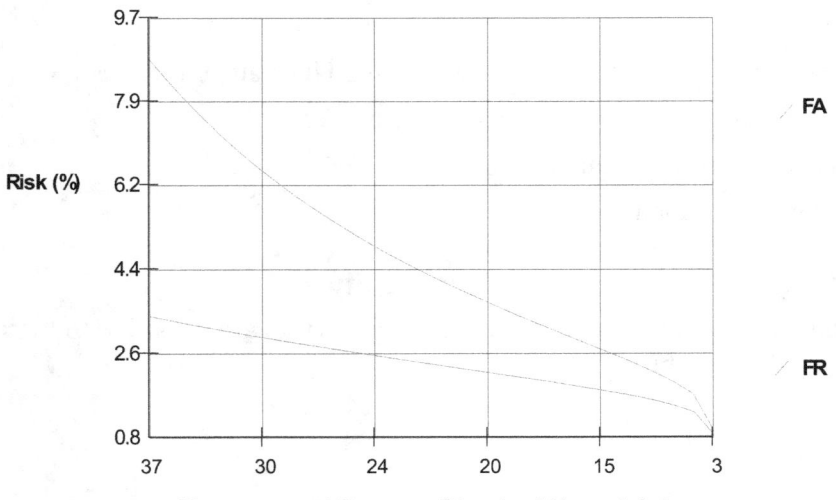

Figure 5-2. Risk vs. Measurement Process Standard Uncertainty.

Shown is a case where the UUT attribute in-tolerance probability is 90%. As the figure indicates, both false accept risk (FA) and false reject risk (FR) are functions of measurement process uncertainty. Note that both false accept risk and false reject risk decrease with decreasing measurement process uncertainty. In the plot, FA corresponds to *UFAR*.

It is of some interest, as Figures 5-1 and 5-2 demonstrate, that risks appear to be more sensitive to UUT attribute in-tolerance probability than to measurement process uncertainty. This is typical of calibration and testing.

The rationale for this can be easily appreciated. For instance, suppose that the UUT attribute in-tolerance probability were 99.999%. In this case, false accept risk would be minimal and fairly insensitive to other variables, simply because there are very few out-of-tolerance attributes to falsely accept in the first place.

These observations are at odds with attempts to control risks with guardbands computed from simple algorithms that take into account measurement process uncertainty relative to UUT attribute specifications. If risks are sensitive to UUT attribute in-tolerance probability, then any control efforts that ignore this variable will produce misleading false accept risk estimates.

5.2.1 Pareto Analysis

Since measurement process uncertainty is a contributing factor to measurement decision risk, this risk may be reduced if the total process uncertainty can be reduced. To maximize the effectiveness of reducing uncertainty, it is beneficial to identify each measurement error and weigh its uncertainty relative to that of other measurement errors.

The tool for performing this evaluation is the Pareto chart, shown in the Figure 5-3. The figure displays the uncertainty breakdown for a typical calibration. In this example, the chart shows that the major contributor is the random error (repeatability) accompanying a sample of measurements made with the UUT attribute. [25]

No.	Error Component	Type	Weight (%)
1	UUT Attribute Repeatability	A	72.432
2	UUT Attribute Resolution	B	15.714
3	Reference Attribute Bias	B	7.022
4	Operator Bias	B	4.783
5	Environment	B	0.049

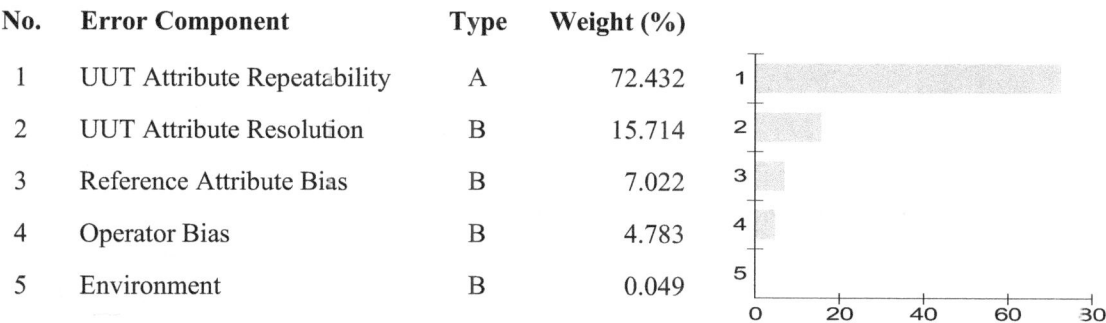

Figure 5-3. Pareto Chart.

Shown are the relative contributions of the uncertainties of individual errors to the uncertainty of the total combined error in a calibration. Clearly, the uncertainty due to repeatability is dominant for the example depicted. This may represent fluctuations due to some ancillary influence. If so, then improving control over the influence in question would seem to be the most effective way of reducing calibration uncertainty.

The repeatability contribution to the total measurement uncertainty may be due to inherent instability in the UUT attribute, fluctuations in the measuring environment, instability in the reference attribute, careless handling of equipment, etc. To reduce repeatability uncertainty, a more stable environment, measuring instrument or more careful equipment handling might be implemented. In general, the specific corrective actions to be taken depend on their effectiveness in controlling the root causes of the dominant errors.

5.2.2 Multiple Independent Measurements

Let δ_1, δ_2, $\cdots$, δ_n represent measurements of a given UUT attribute value using n independent measurement references. Let ε_1, ε_2, ..., ε_n be the measurement biases of the reference attributes, and let u_1, u_2, u_n be the uncertainties in these biases. Then, if the true value of the attribute being measured is x_{true}, the average of measurements made with these attributes is given by

$$\delta = \frac{1}{n}\sum_{i=1}^{n}\delta_i$$

$$= \frac{1}{n}\sum_{i=1}^{n}(e_{UUT,b} + \varepsilon_i)$$

$$= e_{UUT,b} + \frac{1}{n}\sum_{i=1}^{n}\varepsilon_i.$$

---

[25] This would be the case with Calibration Scenario 2, described in Section A.7.2.

If biases are statistically independent, then, using the variance addition rule, we can write the variance in δ as

$$u_{cal}^2 = \text{var}(\delta)$$

$$= \frac{1}{n^2} \sum_{i=1}^{n} \text{var}(\varepsilon_i)$$

$$= \frac{1}{n^2} \sum_{i=1}^{n} u_i^2 .$$

If $u_1 = u_2 = \cdots u_n = u$, then we have the interesting result that

$$u_y = \frac{1}{n}\sqrt{nu^2} = \frac{u}{\sqrt{n}} . \tag{5-1}$$

So, by taking measurements of a quantity δ with n independent measuring devices, each with equal uncertainty, we may reduce the overall bias uncertainty by a factor of $\sqrt{n}$. If the bias uncertainties are not equal, the overall uncertainty is

$$u_y = \frac{1}{n}\sqrt{\sum_{i=1}^{n} u_i^2} . \tag{5-2}$$

5.2.3   Sequential Testing

Sequential testing involves passing a UUT attribute through some number of independent tests or calibrations.[26] In each test, a determination is made as to whether the UUT attribute is in-tolerance. If it is found out-of-tolerance at any stage in the sequence, it is rejected. As can be easily appreciated, the procedure reduces false accept risk by increasing the opportunities for rejecting the attribute. As can also be appreciated, however, is that the procedure *increases* the probability for a false reject. The basic sequential testing process is shown in Figure 5-4. In Figure 5-4, the variable x can represent either the UUT attribute value or bias. The function $f_0(x)$ is the pdf for this variable prior to testing. The pdf $f_1(x)$ is the post-test pdf for UUT attribute values or biases emerging from Test 1. Similar designations apply to successive pdfs. The result of the testing process is the pdf $f_n(x)$.

Analysis of false accept and false reject risks in sequential testing is a complicated process involving considerable computer CPU time. It begins with the determination of $f_1(x)$ using the methods described in Handbook Chapter 2 and Annex 1. This pdf is then used to create a table of values that describe the pre-test UUT attribute distribution for Test 2. Of course to be accurate, the table needs to quasi-continuous. Tables with up to 10,000 entries are not uncommon. This table building effort is made at each step of the process. In most cases, by the time the second or third test in the sequence is reached, the false accept risk becomes negligible.

---

[26] Tests are independent if the value obtained by one measurement does not influence the value obtained by a subsequent measurement. Ensuring complete independence is often impractical, since each test would require the use of a random sample of measurement references of a given type, each calibrated by different independent agencies and handled by a random sample of operators. However, false accept risk can be usually be reduced if different measurement references can be employed for each test, even if they are calibrated by the same organization and used by the same operator.

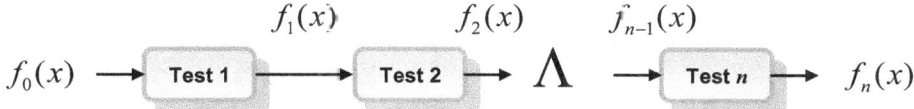

Figure 5-4. Sequential Testing involving *n* Test Steps.

The input to the process is the untested pdf of a UUT attribute. The pdf that emerges from Test 1 may apply to only attributes that pass the test or may apply to a combination of attributes passing the test and attributes that failed the test and have been subsequently adjusted or otherwise corrected. The process continues, culminating with the "post-test" pdf.

Once the final pdf is established, the false accept risk becomes

$$CFAR = 1 - \int_{-L_1}^{L_2} f_n(x)dx .$$

(5-3)

The *CFAR* designation is used because the result applies to the accepted lot of attributes.

5.3   Using Guardbands

As has been shown in Section 5.1 and 5.2, false accept risk can be reduced by increasing the UUT attribute and measurement reference attribute in-tolerance probabilities, reducing uncertainties due to major contributors to measurement error, performing multiple independent measurements and sequential testing.

In some cases, none of the above measures will be practical or even possible. If so, then it may be prudent to fall back on the use of test guardbands.

5.3.1   Guardband Multipliers

It is often useful to relate the range of values $\mathcal{A}$, corresponding to acceptance without correction, to the range of in-tolerance values L using **guardband multipliers**. Let g_1 and g_2 be lower and upper guardband multipliers, respectively. If $-L_1$ and L_2 are the lower and upper UUT attribute tolerance limits, and $-A_1$ and A_2 the corresponding acceptance limits, then

$$A_1 = g_1 L_1$$
$$A_2 = g_2 L_2 .$$

(5-4)

Suppose that g_1 and g_2 are both < 1. Then $\mathcal{A}$ is smaller than L. If guardband multipliers were not employed, then $\mathcal{A}$ and L are the same and *CFAR*, for instance, could be written

$$CFAR_L = 1 - \frac{P(x \in L, y \in L)}{P(y \in L)} .$$

If $\mathcal{A}$ and L are the not same, then *CFAR* could be written

$$CFAR_A = 1 - \frac{P(x \in L, y \in \mathcal{A})}{P(y \in \mathcal{A})} .$$

Now, if $\mathcal{A}$ is smaller than L, then

$$P(x \in \mathsf{L}, y \in \mathcal{A}) < P(x \in \mathsf{L}, y \in \mathsf{L}),$$

and

$$P(y \in \mathcal{A}) < P(y \in \mathsf{L}).$$

It is easy to see that, because $P(y \in \mathcal{A}) / P(y \in \mathsf{L})$ is less than $P(x \in \mathsf{L}, y \in \mathcal{A}) / P(x \in \mathsf{L}, y \in \mathsf{L})$, we have

$$\frac{P(x \in \mathsf{L}, y \in \mathsf{L})}{P(y \in \mathsf{L})} > \frac{P(x \in \mathsf{L}, y \in \mathcal{A})}{P(y \in \mathcal{A})},$$

so that

$$CFAR_A < CFAR_L.$$

It can also be shown that, if $\mathcal{A}$ is smaller than L, then

$$UFAR_A < UFAR_L.$$

Likewise, it can be shown that

$$FRR_A > FRR_L.$$

So, the use of guardbands that set test limits inside tolerance limits reduces false accept risk and increases false reject risk. Conversely, using guardbands that set test limits outside tolerance limits increases false accept risk and reduces false reject risk. The former are called **test guardband limits** and the latter are sometimes referred to as **reporting guardband limits**. Reporting guardband limits are discussed in Section 5.3.3.

5.3.2   Test Guardband Limits

As we have already seen, guardbands are usually specified in terms of a *guardband multiplier*. For example, if the guardband multiplier is 0.9, the test guardband limits are set at 90% of the UUT tolerance limits. Test guardband limits are shown in Figure 5-5.

5.3.2.1   Setting Guardband Multipliers

Test guardband limits are established by setting the guardband multipliers g_1 and g_2 to be less than one. Guardband multipliers are usually symmetric for two-sided tolerance tests or calibrations, but need not be so. Asymmetric multipliers are used when the consequences of accepting an attribute that is out-of-tolerance in one direction are more serious than accepting an attribute that is out-of-tolerance in the other direction. An example is the cannonball discussed earlier: The consequences of a cannonball being bigger than its upper tolerance limit are certainly different from those for a cannonball being smaller that its lower limit.

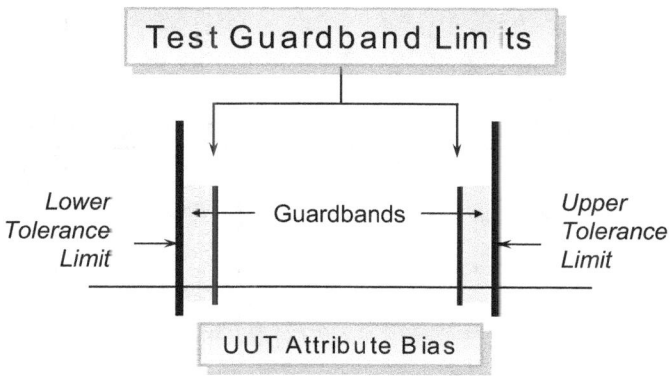

Figure 5-5. Test Guardband Limits.

Limits that trigger an adjustment or other corrective action. Limits are set inside the UUT attribute tolerance limits to reduce false accept risk. Use of these limits also increases false reject risk.

The procedure for setting g_1 and g_2 to achieve an acceptable level of false accept risk is given in Appendix B.

5.3.2.2 Keying Guardbands to Measurement Uncertainty

A proposed method of controlling false accept risk sets test limits inside tolerance limits by some multiple of the calibration uncertainty. This does not explicitly satisfy any false accept risk requirements of a given test or measurement and is not cognizant of the economics surrounding the test or measurement process. In addition, as we have seen, a major contributor to measurement decision risk is the in-tolerance probability or bias uncertainty of the UUT attribute. Since this uncertainty is not included in the calibration uncertainty, any practice that omits it does not qualify as a viable risk control method.

If it is desired to set guardband limits using multiples of the calibration uncertainty, it is possible to link the uncertainty multiple to an allowable false accept risk. This is demonstrated in Table 5-1. Table 5-1 shows the guardband multipliers that would be used to ensure a maximum false accept risk of 1% for various measurement process uncertainties. In the table, "uncertainty ratio"[27] is the ratio of the UUT bias standard uncertainty to the measurement process standard uncertainty.

Table 5-1. Uncertainty k-Factors for a 1% False Accept Risk.

UUT Tolerance Limits:	±100 um
UUT % In-Tol:	95
UUT Bias Distribution:	Normal
Measurement Process Error Distribution:	Normal
UUT Bias Uncertainty	51.0214 um
Maximum Allowable False Accept Risk:	1.0%

---

[27] Not to be confused with "test uncertainty ratio" or TUR, defined in Z540.3 [B-5]. See also Section 1.4 "Terms and Definitions."

Risk Option:				*UFAR*
Process Uncertainty (um)	**Uncertainty Ratio**	**Guardband Multiplier**	**False Reject Risk (%)**	**k-Factor**
0.5102	100:1	1.0479019	< 0.0001	-9.38858826
1.0204	50:1	1.0480591	< 0.0001	-4.70970336
5.1021	10:1	1.0464835	0.1262	-0.91105969
10.2043	5:1	1.0285771	0.8061	-0.28005016
25.5107	2:1	0.9339747	6.1568	0.25881409
51.0214	1:1	0.7391005	26.5683	0.51135339

The **k-factor** is the multiple of the measurement process uncertainty that is subtracted from the 100 um tolerance limit to produce the guardband limits. Notice that, for low uncertainties, the k-factors are negative. This corresponds to setting the test guardband limits *outside* the tolerance limits, i.e., to using guardband multipliers greater than 1. Note also the impact on false reject risk.

Table 5-2 shows the same measurement scenarios as Table 5-1 with the k-factor arbitrarily set at 2 for all cases. Notice the low false accept risks for lower uncertainties. Notice also the accompanying high false reject risks. Obviously, if the situations shown represent cases where a 1% false accept risk is tolerable, for instance, then fixing the k-factor at 2 leads to unnecessary rework expense, especially for uncertainty ratios of around 5 or less. Moreover, if a 1% false accept risk is acceptable, then the cost of this rework is incurred with no appreciable return on investment.

Table 5-2. Risks Associated with a k-Factor of 2.

UUT Tolerance Limits:	±100 um
UUT % In-Tol:	95
UUT Bias Distribution:	Normal
Measurement Process Error Distribution:	Normal
UUT Bias Uncertainty	51.0214 um
k Factor:	2.0
Risk Option:	*UFAR*

Process Uncertainty (um)	**Uncertainty Ratio**	**Guardband Multiplier**	**False Accept Risk (%)**	**False Reject Risk (%)**
0.5102	100:1	0.989796	0.00106	0.2406
1.0204	50:1	0.979591	0.00209	0.4931
5.1021	10:1	0.897957	0.00101	3.0000
10.2043	5:1	0.795914	0.00110	7.6273
25.5107	2:1	0.489786	0.00598	34.0917
51.0214	1:1	N/A	N/A	N/A

Of course other scenarios are possible. It is beyond the scope of this Handbook Annex to cover even a representative selection. It is interesting to note, however, that at uncertainty ratios of 5:1 or less, the false reject risks obtained with a guardband multiplier of 2 may exceed 7.6%.

5.3.3 Reporting Guardband Limits

5.3.3.1 Compensating for Errors in Observed In-tolerance Probabilities

Typically, testing and calibration are performed with safeguards that cause false accept risks to be lower than false reject risks. This is characteristic, for example, of calibration or test equipment inventories with pre-test in-tolerance probabilities higher than 50%. The upshot of this is that, due to the imbalance between false accept and false reject risks, the perceived or *observed* percent in-tolerance will be lower than the actual or *true* percent in-tolerance. This was first reported by Ferling in 1984 as the "True vs. Reported" problem [7]. The issue is discussed at length in Appendix E.

As will be argued in the next section, this discrepancy can have serious repercussions in setting test or calibration intervals. Since these intervals are major cost drivers, the True vs. Reported problem should not be taken lightly.

Through the judicious use of guardbands, the observed percent in-tolerance can be brought in line with the true in-tolerance percentage. With pre-test in-tolerance probabilities higher than 50%, this usually means setting test guardband limits outside the tolerance limits.

5.3.3.2 Implications for Interval Analysis

Ordinarily, intervals are set to achieve end-of-period (EOP) in-tolerance levels of around 80% to 95%. These levels are referred to as **reliability targets**.[28] If intervals are analyzed using test or calibration history, and high reliability targets are employed, the intervals ensuing from the analysis process can be seriously impacted by reported out-of-tolerances. In other words, with high reliability targets, only a few reported out-of-tolerances can result in drastically shortened intervals.

Since this is the case, and, since the length of test or calibration intervals is a major cost driver, it is prudent to ensure that perceived out-of-tolerances not be the result of false reject risk. This is one of the central reasons why striving for reductions in false accept risk must be made with caution, since reductions in false accept risk cause increases in false reject risk. At the very least, attempts to control false accept risk should be made with cognizance of the return on investment and an understanding of the trade-off in increased false reject risk and shortened calibration intervals.

Reporting guardband limits are used to ameliorate the cost of periodic calibration while maintaining the desired reliability target. This is done by achieving a reported EOP percent in-tolerance that is equal to the true EOP percent in-tolerance. Hence, a UUT attribute would be *reported* as out-of-tolerance only if its value fell outside its reporting guardband limits. The methodology for establishing such limits is presented in Appendix E.

---

[28] Setting risk-based reliability targets is discussed in Appendix H.

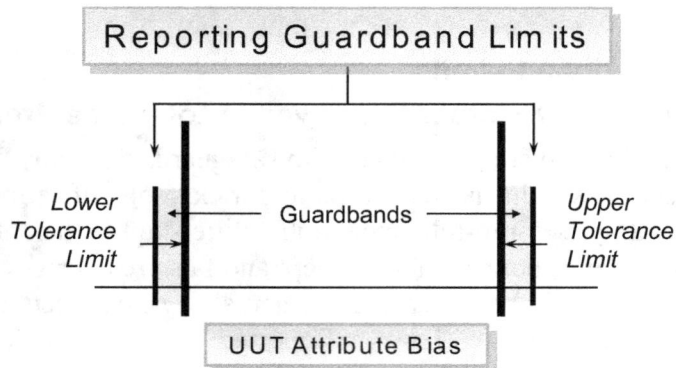

Figure 5-6. Reporting Guardband Limits.

Limits used to report out-of-tolerances. Limits are set outside the UUT attribute tolerance limits to equalize false accept risk and false reject risk. This has the effect of adjusting the reported percent out-of-tolerance to match the true percent out-of-tolerance.

5.3.3.3 Summary

To accommodate both the need for low false accept risks and accurate in-tolerance reporting, it is required that two sets of guardbands be employed. One, ordinarily set inside the tolerances, would apply to withholding items from use or to triggering attribute adjustment actions. The other, ordinarily set outside the tolerances, would apply to in- or out-of-tolerance reporting for purposes of calibration interval analysis and calibration feedback reporting.

Test Guardband limits

The first set of guardband limits is called test guardband limits. Test guardband limits are those that are normally thought of when guardbands are discussed. Test guardbands are used to control false accept risks.

> **Test guardband limits trigger adjustments or other corrective actions.**

Reporting Guardband limits

Reporting guardband limits are used to compensate for the True vs. Reported problem. An attribute would be *reported* as out-of-tolerance only if its value fell outside its reporting guardband limits.

> **Reporting guardband limits comprise pass-fail criteria for reporting out-of-tolerances.**

5.4 Bayesian Guardbands

Although Bayesian analysis is ideal for a bench-level application, it can also be used to develop process-level guardband limits. Thus, if we are constrained by a maximum acceptable false accept risk, i.e., a minimum acceptable P_{in}, we can solve for a maximum acceptable estimated UUT attribute bias δ_c. This maximum acceptable estimate comprises the guardband limit.

From Eq. (4-32), the UUT attribute in-tolerance probability, given a calibration result δ, is given by

$$P_{in} = \Phi\left(\frac{L_1 + \beta}{u_\beta}\right) + \Phi\left(\frac{L_2 - \beta}{u_\beta}\right) - 1$$

where β and u_β are given in Eqs. (4-24) and (4-25), respectively.

The solution process attempts to find a value of $\beta = \beta_c$ such that P_{in} is equal to some minimum allowable value P_c. The bisection method described in Appendix F has been found to be useful for this. The root to be solved for is

$$F = \Phi\left(\frac{L_1 + \beta_c}{u_\beta}\right) + \Phi\left(\frac{L_2 - \beta_c}{u_\beta}\right) - 1 - P_c. \qquad (5\text{-}5)$$

Once the value of β_c has been found for which $F = 0$, the guardband limit δ_c is computed from

$$\delta_c = \frac{u_A^2}{u_{UUT,b}^2}\beta_c, \qquad (5\text{-}6)$$

where u_A and $u_{UUT,b}$ are defined in Section 4.3.4. For cases where the UUT attribute tolerance limits are single-sided upper or single-sided lower, L_1 or L_2 is set to a physical limiting value. In some cases this value is essentially infinite. For instance, if the tolerance limit is single-sided upper with a lower limit of $-\infty$, Eq. (5-5) becomes

$$F = \Phi\left(\frac{L_2 - \beta_c}{u_\beta}\right) - P_c,$$

and, if the tolerance limit is single-sided lower with an upper limit of $+\infty$,

$$F = \Phi\left(\frac{L_1 + \beta_c}{u_\beta}\right) - P_c.$$

Table 5-3 Shows Bayesian guardband limits for a UUT attribute calibration with an uncertainty ratio of 4 1, and different minimum EOP percent in-tolerance criteria. In the table, the tolerance limits are two-sided symmetric and the UUT attribute bias is assumed to be normally distributed.

Table 5-3. Bayesian Guardband Limits.

UUT Attribute Tolerance Limits	± 10 mV
UUT Attribute EOP Percent In-Tolerance	95.00
UUT Attribute a priori Bias Uncertainty $u_{UUT,b}$	5.1021 mV
Calibration Uncertainty u_{cal}	1.2755 mV
Calibration 95% Expanded Uncertainty U_{95}	2.5 mV

Max Allowable Risk (%)	± Guardband Limits (mV)
0.10	6.5621
0.50	7.2384
1.00	7.5664
2.00	7.9248
3.00	8.1522
4.00	8.3233
5.00	8.4624

5.5   Confidence Level Guardbands

Like Bayesian analysis, confidence level analysis is typically applied as a bench-level method. However, the method can also be implemented with process-level guardband limits. Just as with Bayesian guardbands, if we are constrained by a minimum acceptable P_{in}, we can solve for a maximum acceptable estimated UUT attribute bias δ_c. This maximum acceptable estimate comprises the guardband limit.

From Eq. (4-35), the UUT attribute in-tolerance probability, given a calibration result δ, is given by

$$P_{in} = \Phi\left(\frac{L_1+\delta}{u_{cal}}\right) + \Phi\left(\frac{L_2-\delta}{u_{cal}}\right) - 1. \tag{5-7}$$

where u_{cal} is defined in Table 4-7.

The solution process attempts to find a value of $\delta = \delta_c$ such that P_{in} is a minimum allowable value P_c. The bisection method described in Appendix F can be used for this. The root to be solved for is

$$F = \Phi\left(\frac{L_1+\delta_c}{u_{cal}}\right) + \Phi\left(\frac{L_2-\delta_c}{u_{cal}}\right) - 1 - P_c. \tag{5-8}$$

Once the value of δ_c has been found for which $F = 0$, we have the appropriate guardband limit. For cases where the UUT attributed tolerance limits are single-sided upper or single-sided lower, L_1 or L_2 are set to ∞, i.e.,

$$F = \Phi\left(\frac{L_2-\delta_c}{u_{cal}}\right) - P_c,$$

if the tolerance limit is single-sided upper, and

$$F = \Phi\left(\frac{L_1+\delta_c}{u_{cal}}\right) - P_c.$$

if the tolerance limit is single-sided lower.

Table 5-4 Shows confidence level guardband limits for a UUT attribute calibration with a nominal 4:1 TUR, as defined in Section 3.5.1, and different minimum EOP percent in-tolerance criteria. In the table, the tolerance limits are two-sided symmetric and the UUT attribute bias is assumed to be normally distributed.

Table 5-4. Confidence Level Guardband Limits.

UUT Attribute Tolerance Limits	±10 mV
UUT Attribute EOP Percent In-Tolerance	95.00
UUT Attribute a priori Bias Uncertainty $u_{UUT,b}$	5.1021
Calibration Uncertainty u_{cal}	1.2755 mV
Calibration 95% Expanded Uncertainty U_{95}	2.5 mV

Max Allowable Risk (%)	± Guardband Limits (mV)
0.10	6.0584
0.50	6.7145
1.00	7.0327
2.00	7.3804
3.00	7.6010
4.00	7.7670
5.00	7.9020

Notice that a comparison of Table 5-4 with 5-3 shows that guardband limits with the confidence level method are consistently tighter than with the Bayesian method.

5.6 Minimizing Costs

Guardbands may be used to establish a compromise between false accept risks and false reject risks. If the cost of a false reject is prohibitive, for example, it may be desired to set test guardband limits that reduce false reject risk at the expense of increasing false accept risk. If, on the other hand, the cost of false accepts is prohibitive, it may be desired to reduce false accept risk at the expense of increasing false reject risk.

A simplified cost modeling approach is described below that balances false accept risk and false reject risk to optimize total cost.

A more comprehensive end-to-end approach that takes into account equipment life cycle costs, calibration and testing support costs, and the cost of undesired outcomes is described Handbook Annex 1 and in References [13] and [36].

5.6.1 A Simplified Model

If it is not feasible to perform a detailed cost analysis, as described in Handbook Chapter 2 and Annex 1, it may be possible to implement a simplified approach which focuses primarily on the cost of false rejects and false accepts expressed in terms of averages.

5.6.1.1 False Reject Cost

With regard to the cost of a false reject, a good start would be to develop an average labor and parts estimate for recalibration, rework or other corrective action for items with a specific attribute of interest that is functioning within its tolerance limits. Letting $\overline{C}_{renew,G}$ represent this average, the annual cost of false rejects for a particular type of item would then be given by

$$C_{FR} = \frac{N_{UUT}\overline{C}_{renew,G}FRR}{I},$$

where N_{UUT} is the number of items in inventory with the specific attribute under consideration, FRR is the false reject risk associated with testing or calibration, and I is the test or calibration interval and the G ("good") subscript indicates that the cost corresponds to renewing in-tolerance attributes. For an end item, the quotient N_{UUT}/I may represent the number of items tested per year.

Note that the above cost does not include the cost of increased frequency of calibration due to shortening of intervals in response to false rejects, nor does it include the cost of generating unnecessary out-of-tolerance reports or other clerical actions.

5.6.1.2 False Accept Cost

The cost of a false accept is more difficult to nail down without resorting to a fairly complicated cost/utility model. As mentioned earlier, the cost of a false accept is felt in terms of negative outcomes resulting from the use of an out-of-tolerance attribute. Table 5-5 describes the variables used in the simplified model.

Table 5-5. Variables Used in the Simplified Cost Model.

Variable		Description
$e_{UUT,b}$	-	the bias of the tested or calibrated attribute
n_E	-	the number of possible situations in which the use of the attribute of a tested or calibrated item may result in a negative outcome
$C_i(e_{UUT,b})$	-	the cost of a negative outcome of the ith possible usage event, $i = 1, 2, \cdots, n_E$
$f(e_{UUT,b})$	-	the EOP probability distribution function for $e_{UUT,b}$
P_i	-	the probability that the ith event will occur

With these variables, we can estimate an average performance cost of a nonconforming attribute according to

$$\overline{C}_{perf,B} = \sum_{i=1}^{n_E} P_i \left[1 - \int_L f(e_{UUT,b})C_i(e_{UUT,b})de_{UUT,b} \right],$$

where L is the performance region for $e_{UUT,b}$, as defined before, and the B ("bad") subscript indicates that the cost corresponds to out-of-tolerance attributes.

With the above expression, the annual cost of falsely accepting the attribute of interest may be written

$$C_{FA} = \frac{N_{UUT}\overline{C}_{perf,B}FAR}{I}.$$

where FAR may be $UFAR$ or $CFAR$, as appropriate.

5.6.1.3 Total Cost

The total annual cost due to measurement decision risk is just the sum of C_{FR} and C_{FA}:

$$C_{risk} = C_{FR} + C_{FA}.$$

This cost needs to be added to the annual support cost for testing or calibration of the UUT attribute. This cost may be given by

$$C_{ts} = \frac{N_{MTE} \bar{C}_{serv}}{I_{MTE}},$$

where

N_{MTE} - the number of items in inventory that are used to calibrate or test the UUT attribute

$\bar{C}_{serv}$ - the average cost of test or calibration service for the MTE

I_{MTE} - the interval for the MTE.

The total cost that comprises the management variable to be minimized is

$$C_{total} = C_{ts} + C_{risk}.$$

5.6.1.4 Optimizing Risk Management

In optimizing risk, *FRR* and *FAR* are adjusted in such a way that C_{total} is minimized. This may be done by experimentation with guardbands, in-tolerance percentages, substituted equipment, multiple independent measurement or sequential testing schemes.[29]

Experimentation with Guardbands

In experimenting with guardbands, the variable C_{ts} is held fixed, as are all variables in C_{FA} and C_{FR}, with the exception of *FAR* and *FRR*. In other words, the only variables that need to be varied are *FAR* and *FRR*. This is done by moving guardband multipliers in and out in a "hunt and peck" process.

Experimentation with In-Tolerance Percentages

Changing the in-tolerance probabilities of the UUT and reference attributes will result in changes to *FAR* and *FRR* that may produce the desired minimization of C_{total}. However, if the in-tolerance probability for the tested or calibrated attribute is changed, the variable *I* must also be updated in the above expressions. If the in-tolerance probability for the reference attribute is changed, the variable I_{MTE} must also be modified.

Substituting Equipment

Modifying the basic accuracy of the reference attribute may impact C_{total} in the desired way. However, if this is done, the variables I_{MTE} and $\bar{C}_{serv}$ may also need to be updated.

---

[29] A more comprehensive cost analysis is described in Handbook Chapter 2 and Annex 1 in which C_{total} includes total life cycle costs and the cost of employing a tested end item. The latter includes such considerations as criticality of use and the effect of end item utility on expenses incurred or avoided in its use.

Multiple Independent Measurements / Sequential Testing

As discussed in Section 5.2.3, false accept risk may be reduced through performing multiple independent measurements during calibration or testing or by implementing a sequential testing procedure. This reduction comes at a price in that (1) such schemes are more costly than single measurement schemes and (2) the reduction in false accept risk is accompanied by an increase in false reject risk.

References

[1] Eagle, A., "A Method for Handling Errors in Testing and Measuring," *Industrial Quality Control*, March, 1954.

[2] Grubbs, F. and Coon, H., "On Setting Test Limits Relative to Specification Limits," *Industrial Quality Control*, March, 1954.

[3] Hayes, J., "Factors Affecting Measuring Reliability," U.S. Naval Ordnance Laboratory, TM No. 63-106, 24 October 1955.

[4] Castrup, H., *Evaluation of Customer and Manufacturer Risk vs. Acceptance Test In-Tolerance Level*, TRW Technical Report No. 99900-7871-RU-00, April 1978.

[5] Ferling, J., *Calibration Interval Analysis Model*, SAI Comsystems Technical Report, Prepared for the U.S. Navy Metrology Engineering Center, Contract N00123-87-C-0589, April 1979.

[6] Kuskey, K., *Cost-Benefit Model for Analysis of Calibration-System Designs in the Case of Random-Walk Equipment Behavior*, SAI Comsystems Technical Report, Prepared for the U.S. Navy Metrology Engineering Center, Contract N00123-76-C-0589, January 1979.

[7] Ferling, J., "The Role of Accuracy Ratios in Test and Measurement Processes," *Proc. Measurement Science Conference*, Long Beach, January 1984.

[8] Ferling, J. and Caldwell, D., "Analytic Modeling for Electronic Test Equipment Adjustment Policies," *Proc. NCSL Workshop and Symposium*, Denver, July 1989.

[9] Weber, S. and Hillstrom, A., *Economic Model of Calibration Improvements for Automatic Test Equipment*, NBS Special Publication 673, April 1984.

[10] Deaver, D., "How to Maintain Your Confidence," *Proc. NCSL Workshop and Symposium*, Albuquerque, July 1993.

[11] Deaver, D., "Guardbanding with Confidence," *Proc. NCSL Workshop and Symposium*, Chicago, July - August 1994.

[12] BIPM, *International vocabulary of metrology — Basic and general concepts and associated terms* (VIM), 3rd Ed., JCGM 200:2008.

[13] Castrup, H., "Calibration Requirements Analysis System," *Proc. NCSL Workshop and Symposium*, Denver, July 1989.

[14] Castrup, H., "Uncertainty Analysis for Risk Management," *Proc. Measurement Science Conference*, Anaheim, January 1995.

[15] Castrup, H., "Analyzing Uncertainty for Risk Management," *Proc. 49th ASQC Annual Quality Congress*, Cincinnati, May 1995.

[16] Castrup, H., "Uncertainty Analysis and Attribute Tolerancing," *Proc. NCSL Workshop and Symposium*, Dallas, 1995.

[17] Castrup, H., "Risk-Based Control Limits," *Proc. Measurement Science Conference*, Anaheim, January 2001.

[18] Castrup, H., "Test and Calibration Metrics," Note to the NCSLI Metrology Practices Committee, 24 February 2001.

[19] Castrup, H., "Risk Based Control Limits," *Proc. Measurement Sciences Conference*, Anaheim, January 2001.

[20] Castrup, H., "Test and Calibration Metrics," Note to the NCSLI Metrology Practices Committee, 24 February 2001.

[21] Castrup, H., "Risk Analysis Methods for Complying with Z540.3," *Proc. NCSLI Workshop & Symposium*, St. Paul, 2007.

[22] Castrup, H., "Decision Risk Analysis for Alternative Calibration Scenarios," *Proc. NCSLI Workshop & Symposium*, Orlando, 2008.

[23] Jackson, D., "Measurement Risk Analysis Methods," *Proc. Meas. Sci. Conf.*, Anaheim, 2005.

[24] Jackson, D., "Measurement Risk Analysis Methods with Multiple Test Points," *Proc. Meas. Sci. Conf.*, Anaheim, 2006.

[25] Mimbs, S., "Measurement Decision Risk – The Importance of Definitions," *Proc. NCSLI Workshop & Symposium*, St. Paul, 2007.

[26] NCSLI, *Determining and Reporting Measurement Uncertainties*, Recommended Practice RP-12, NCSL International, Under Revision.

[27] NASA, *Measurement Uncertainty Analysis Principles and Methods*, NASA-HDBK-8730.19-3, National Aeronautics and Space Administration, 2009.

[28] Castrup, H., Intercomparison of Standards: General Case, SAI Comsystems Technical Report, U.S. Navy Contract N00123-83-D-0015, Delivery Order 4M03, March 16, 1984.

[29] Jackson, D., *Instrument Intercomparison: A General Methodology*, Analytical Metrology Note AMN 86-1, U.S. Navy Metrology Engineering Center, NWS Seal Beach, January 1, 1986; and "Instrument Intercomparison and Calibration," Proc. 1987 MSC, Irvine, January 29, 30.

[30] Castrup, H., "Analytical Metrology SPC Methods for ATE Implementation," *Proc. NCSL Workshop & Symposium*, Albuquerque, August 1991.

[31] MIL-STD 45662A, *Calibration Systems Requirements*, U.S. Dept. of Defense, 1 August 1988.

[32] ANSI/NCSL Z540-1:1994, *Calibration Laboratories and Measuring and Test Equipment – General Requirements*, July 1994.

[33] ANSI/NCSL Z540.3-2006, *Requirements for the Calibration of Measuring and Test Equipment*, August 2006.

[34] Bayes, T., "An Essay towards solving a Problem in the Doctrine of Chances," *Phil. Trans.* **53**, 1763, 376-98.

[35] Castrup, S., "Interpreting and Applying Equipment Specifications," Presented at the NCSLI International Workshop and Symposium, Washington, DC, August, 2005. Revised 5/31/08. Available at www.isgmax.com.

[36] Castrup, H., "Applying Measurement Science to Ensure End Item Performance," *Proc. Meas. Sci. Conf.*, Anaheim, 2008.

[37] Castrup, H., "Measurement Quality Assurance Principles and Applications," Presented at the NASA MCWG Meeting, Cocoa Beach, February 2008.

[38] NCSL, Recommended Practice RP-1, *Establishment and Adjustment of Calibration Intervals*, January 1996.

[39] NCSL, Recommended Practice RP-5, *Measurement and Test Equipment Specifications*, In Revision.

[40] NASA, NASA-HDBK-8730.19, *Measurement Quality Assurance Handbook*, National Aeronautics and Space Administration, 2008.

[41] Welch, B., *Biometrica* **29**, 1947.

[42] Satterthwaite, F., *Psychometrica* **6**, 1946.

[43] AccuracyRatio 1.6, © 2004, Integrated Sciences Group, www.isgmax.com.

[44] ISO/IEC 17025 1999(e), General Requirements for the Competence of Testing and Calibration Laboratories, ISO/IEC, December 1999.

[45] JPL, *Metrology – Calibration and Measurement Processes Guidelines*, NASA Reference Publication 1342, June 1994.

Appendix A - Measurement Uncertainty Analysis

This appendix describes principles and methods of measurement uncertainty analysis. A systematic process for computing and combining uncertainty estimates is presented. The process yields total combined uncertainties that facilitate effective analyses of measurement decision risk. The appendix concludes with a discussion of four calibration scenarios for which relevant error sources are identified and specific recipes for combining calibration uncertainties are provided.

> **Note:** The uncertainty analysis methodology documented in this appendix is condensed from uncertainty analysis methods and principles presented in NCSLI's Recommended Practice RP-12 [A-1]. For a complete treatment of the subject, including analysis procedures and examples, the reader is referred to this document.

A.1 Appendix A Nomenclature

Nomenclature used in this Appendix for the principal quantities is summarized in Table A-1. The notation for other quantities can be determined by applying the notation of Table A-2.

Table A-1. Appendix A Nomenclature.

Quantity	Description
UUT	Unit Under Test. The device or artifact undergoing calibration.
Attribute	A measurable property of a device, substance or other quantity.
MTE	Measuring or Test Equipment. A reference standard or item of test equipment used as a measurement reference.
u	Standard uncertainty.
x	A measured value taken by the UUT attribute.
y	A measured value taken by the reference attribute.
μ_x	The mean value of a population of values represented by the variable x. Sometimes taken to represent the true value of x.
ε_x	The error in the measurement of x.
σ_x	The standard deviation of the population of values represented by the variable x. Equated with u_x the uncertainty in x.
var	The variance operator.
cov	The covariance operator.
$\rho(e_{x_1}, e_{x_2})$	The correlation coefficient for correlations between errors e_{x_1} and e_{x_2}.
ν	Degrees of freedom.
$t_{\alpha,\nu}$	The t-statistic for a confidence level of $1 - \alpha$ and degrees of freedom ν.
ε_m	The total error in the measurement of the value of an attribute.
$e_{UUT,b}$	(1) The bias of a UUT attribute as received for calibration. (2) The quantity estimated by UUT calibration.
$u_{UUT,b}$	The uncertainty in the bias of a UUT attribute as received for calibration. Equal to the standard deviation of the $e_{UUT,b}$ distribution.
$e_{MTE,b}$	The bias of the MTE attribute used to calibrate the UUT attribute.
$\varepsilon_{UUT,m}$	The error in measurements made with the UUT attribute or the error in measuring the UUT attribute's value with a comparator.

Quantity	Description
$\varepsilon_{MTE,m}$	The error in measurements made with the MTE attribute or the error in measuring the MTE attribute's value with a comparator.
δ	The result of a UUT calibration, i.e., an estimate of $e_{UUT,b}$ obtained by calibration.
ε_{cal}	The error in δ.
u_{cal}	The uncertainty in ε_{cal}.
x_n	The nominal value of a UUT attribute.
x_{true}	The true value of a UUT attribute.
y_n	The nominal value of an MTE attribute.
y_{true}	The true value of an MTE attribute.
x_c	The value of the UUT attribute indicated by a measurement taken with a comparator.
$e_{c,b}$	The bias in a comparator indication.
$\Phi(\zeta)$	The cumulative normal distribution function for a random variable with value ζ.

A.2   Uncertainty Analysis Fundamentals

This section provides an overview look at the elements of the uncertainty estimation that comprise the foundation on which measurement decision risk analysis is built. After some preliminary remarks, the relationship between measurement error and equipment attribute biases will be described, followed by a mathematical definition of measurement uncertainty. Error sources will be identified and a systematic method for combining uncertainties due to combinations of error sources will be presented. Methods of computing measurement uncertainty will be outlined and a powerful method for estimating uncertainty for multivariate measurements will be described.

While the subject of uncertainty analysis has been carefully treated in this document, the reader is advised that, as the above note indicates, a full and rigorous discussion can be found in reference [A-1].

A.2.1   Preliminaries

In making uncertainty estimates, we keep in mind a few basic concepts. Chief among these are the following:

- All measurements are accompanied by error.

- Measurement errors are <u>random variables</u>. This means that whenever we make measurements, the errors in these measurements vary with respect to sign and magnitude.

- Errors (and attribute deviations from nominal or "biases") follow <u>probability distributions</u>. The way that errors vary can be described statistically. In statistical descriptions, errors are said to be distributed in such a way that the sign and magnitude of a given error has associated with it a probability of occurrence.

These preliminary observations are summarized in an important uncertainty analysis axiom, stated here as Axiom 1:

In applying this axiom, we keep two important definitions in mind:

Population - All the values that a random variable can attain.

Distribution - A functional relationship between the value of a random variable and the probability of its occurrence.

A.2.2　Error Distributions

How errors combine and their relationships to measurement uncertainties can be better understood by considering the ways in which they are distributed.

Suppose that the attribute being measured, referred to as the *UUT attribute*, has a specified nominal value around which members of its population are distributed. Such distributions may be symmetric about the population mean or *expectation* value, as in the case of normally distributed attribute values, or may be asymmetrical about the nominal value, as with attributes that follow a lognormal distribution.

For purposes of discussion, imagine that the population of a given UUT attribute consists of a particular inventory of items. For example, the inventory may consist of some number of gage blocks of a particular nominal dimension from a particular manufacturer. If a gage block is selected at random from its inventory, it will have a specific true value.

If the gage block is measured by a reference device, also selected randomly from its population, two considerations arise. First, the inventory of reference devices may have a mean systematic offset. Second, the particular device selected may have an additional systematic offset relative to its population mean. The combination of these offsets or biases is called *measurement bias*. Measurement bias can be thought of as the difference between the mean of a sample of measured values, obtained with a given reference attribute, and the true value of the UUT attribute or measurand.

Measurement processes are rarely completely stable. There is nearly always some fluctuation due to the measuring system, the UUT attribute, the measuring technician, or a combination of these variables. These fluctuations appear as random error. Finally, there are errors of perception due to operator bias or to the finiteness of the resolution of the measuring system. These errors combine with the random errors to produce a distribution about the measured mean value.

The combination of measurement bias and other errors constitutes the difference between a measured value and the UUT attribute's true value, i.e., it constitutes the measurement error.

A.2.2.1 Error Sources

Error sources are variables that contribute to measurement error. Error sources are also referred to as *process errors*. The most commonly encountered are

- Measurement Bias - A systematic difference between the value measured by a reference attribute and the true value of the UUT attribute or measurand.

- Random Error (Repeatability) - Errors that are manifested in differences in value from one measurement to the next.

- Resolution Error - The difference between a "sensed" value and the value indicated by a measuring device.

- Digital Sampling Error - Error due to the granularity of digital representations of analog values.

- Computation Error - Error due to computational round-off, regression fit, interpolation or extrapolation.

- Operator Bias (Reproducibility) - Error due to quasi-persistent bias in operator perception and/or technique.

- Stress Response Error - Error caused by response of a calibrated UUT attribute to stress following calibration.

- Environmental/Ancillary Factors Error - Error caused by environmental effects and/or fluctuations in ancillary equipment.

A.2.2.2 Applicable Distributions

Distributions for Type A and Type B analysis are shown below. The Type B estimation procedure has been refined so that standard deviations can be estimated for both normal and non-normal populations and in cases where the confidence limits are asymmetric or even single-sided. For constrained non-normal distributions, the distribution limits are designated $\pm a$ and the tolerance limits are denoted $\pm L$. The following notation is used in expressions for parameters or statistics of the distributions:

p - in-tolerance or "containment" probability
u - distribution standard deviation (standard uncertainty)
Φ - the distribution function for the normal distribution
Φ^{-1} - the inverse distribution function for the normal distribution

The relevant error probability density functions $f(\varepsilon)$ and distribution standard deviations are as follows.

Normal Distribution

Probability Density Function

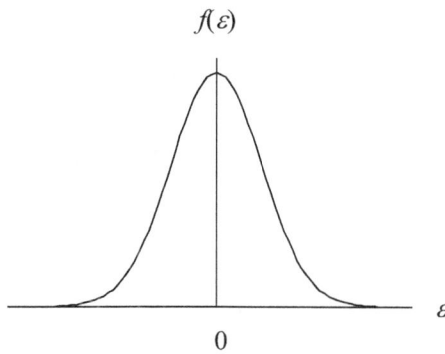

Figure A-1. The Normal Error Distribution.

The "workhorse" distribution used to represent most attribute biases and measurement errors.

$$f(\varepsilon) = \frac{1}{\sqrt{2\pi}u} e^{-\varepsilon^2/2u^2}$$

Distribution Standard Deviation

$$u = \frac{L}{\varphi(p)},$$

where

$$\varphi(p) = \Phi^{-1}\left[(1+p)/2\right]$$

Lognormal Distribution

Probability Density Function

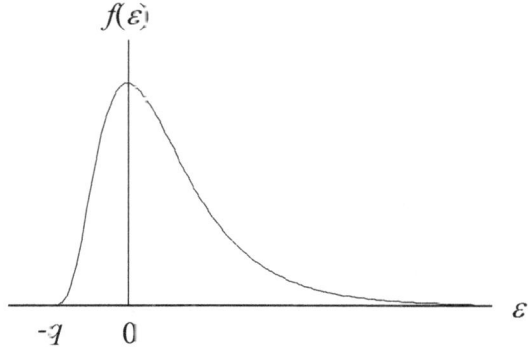

Figure A-2. The Lognormal Distribution.

A useful distribution for attributes with asymmetric tolerance limits. Shown is a "right-handed" lognormal distribution, i.e., one where the lower distribution limit is < 0. Left-handed lognormal distributions, where this limit > 0, are also possible.

$$f(\varepsilon) = \frac{1}{\sqrt{2\pi}\lambda(\varepsilon+q)} \exp\left\{-\left[\ln\left(\frac{\varepsilon+q}{m+q}\right)\right]^2 \Big/ 2\lambda^2\right\}$$

Distribution Parameters

$-q$ - limiting value

m - distribution median

λ - shape parameter

Distribution Standard Deviation

$$u = |m + q| \, e^{\lambda^2/2} \sqrt{e^{\lambda^2} - 1}$$

Uniform Distribution

Probability Density Function

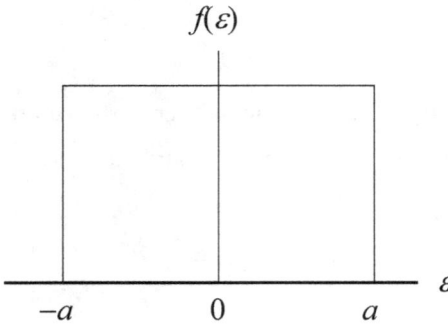

Figure A-3. The Uniform Error Distribution.

A useful distribution for digital resolution error and signal quantization error.

$$f(\varepsilon) = \begin{cases} \dfrac{1}{2a}, & -a \le \varepsilon \le a \\ 0, & \text{otherwise} \end{cases}$$

Distribution Standard Deviation

$$u = \frac{L/p}{\sqrt{3}} = \frac{a}{\sqrt{3}}$$

Triangular Distribution

Probability Density Function

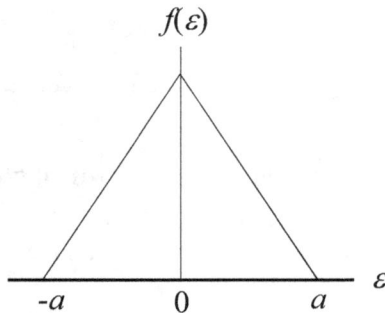

Figure A-4. The Triangular Error Distribution.

The distribution for the sum of two uniformly distributed errors with equal limits and mean values.

$$f(\varepsilon) = \begin{cases} (\varepsilon + a)/a^2, & -a \le \varepsilon \le 0 \\ (a - \varepsilon)/a^2, & 0 \le \varepsilon \le a \\ 0, & \text{otherwise.} \end{cases}$$

Distribution Standard Deviation

$$u = \frac{a}{\sqrt{6}},$$

where

$$a = \frac{L}{1 - \sqrt{1 - p}}, \quad L \le a.$$

Quadratic Distribution

Probability Density Function

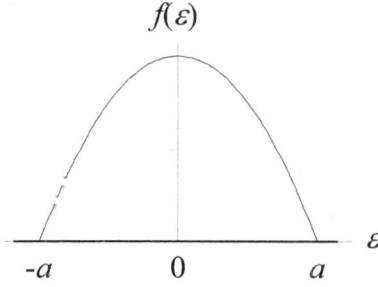

Figure A-5. The Quadratic Error Distribution.

A good distribution for measurement errors or attribute values with a central tendency but a wide spread between distribution limits.

$$f(\varepsilon) = \begin{cases} \dfrac{3}{4a}\left[1 - (\varepsilon/a)^2\right], & -a \le \varepsilon \le a \\ 0, & \text{otherwise.} \end{cases}$$

Distribution Standard Deviation

$$u = \frac{a}{\sqrt{5}},$$

where

$$a = \frac{L}{2p}\left(1 + 2\cos\left[\frac{1}{3}\arccos(1 - 2p^2)\right]\right) \quad -1 < p < 1.$$

Cosine Distribution
Probability Density Function

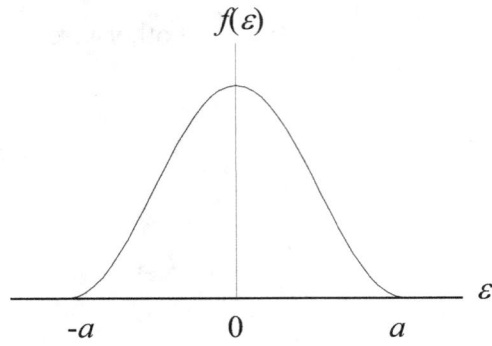

$f(\varepsilon)$

$-a$ 0 a ε

Figure A-6. The Cosine Error Distribution

Describes measurement errors and attribute biases with normal distribution tendencies but with physical bounding limits.

$$f(\varepsilon) = \begin{cases} \dfrac{1}{2a}\left[1 + \cos\left(\dfrac{\pi\varepsilon}{a}\right)\right], & -a \leq \varepsilon \leq a \\ 0, & \text{otherwise}. \end{cases}$$

Distribution Standard Deviation

$$u = \frac{a}{\sqrt{3}}\sqrt{1 - \frac{6}{\pi^2}}\ .$$

The distribution limit a is solved from

$$\frac{1}{\pi}\sin(\pi x) + x - p = 0\,,$$

where

$$x = L\,/\,a.$$

U-Distribution
Probability Density Function

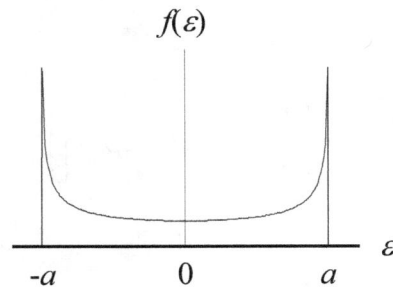

$f(\varepsilon)$

$-a$ 0 a ε

Figure A-7. The U Distribution

The distribution for measurement errors or attribute biases that vary in a periodic sinusoidal manner. Applies to quantities regulated by automated control systems.

$$f(\varepsilon) = \begin{cases} \dfrac{1}{\pi\sqrt{a^2 - \varepsilon^2}}, & -a < x < a \\[2mm] 0, & \text{otherwise.} \end{cases}$$

Distribution Standard Deviation

$$u = \frac{a}{\sqrt{2}},$$

where

$$a = \frac{L}{\sin(\pi p / 2)}.$$

A.2.2.3 Recommendations for Selecting an Error Distribution

The following are offered as guidelines for selecting an appropriate error distribution:

1. Unless information to the contrary is available, the normal distribution should be applied as the default distribution.

2. If it is suspected that the distribution of the value of interest is skewed, apply the lognormal distribution. In using the normal or lognormal distribution, some effort must be made to estimate a containment probability. If a set of containment limits is available, but 100% containment has been observed, then the following is recommended:

   - If the value of interest has been subjected to random usage or handling stress and the resulting error is assumed to possess a central tendency, apply the cosine distribution. If it is suspected that the resulting errors are more evenly distributed, apply either the quadratic or half-cosine distribution, as appropriate. The triangular distribution may be applicable to estimating uncertainty due to interpolation errors, and, under certain circumstances, when dealing with attribute biases following testing or calibration.

   - If the value of interest varies sinusoidally, with random phase, apply the U-distribution.

   - To estimate the uncertainty due to the resolution error of a digital readout, apply the uniform distribution. This distribution is also applicable to estimating the uncertainty due to quantization error and the uncertainty in RF phase angle.

A.2.3 Error and Uncertainty

In the preliminary remarks it was stated that the distributions of measurement errors are used to estimate uncertainty. We now examine this remark in detail and establish the mathematical relationship between error and uncertainty.

A.2.3.1 Statistical Variance

In the previous section, we discussed measurement errors qualitatively. We now proceed to quantify measurement errors in terms of their probability distributions.

A distribution has been successfully specified when we can obtain values for its various statistics. Principal among these is the **variance** of the distribution. The variance of a probability distribution for a variable x is the **mean square error** of the distribution, given by

$$\text{Mean Square Error} = \int_{-\infty}^{\infty} (x - \mu_x)^2 f(x)dx \,,$$

where μ_x is the mean value of the population of values for x and $f(x)$ is the probability density function for x.[30]

$$\text{Mean Square Error } (x) = \int_{-\infty}^{\infty} (x - \mu_x)^2 f(x)dx$$

$$= \text{var}(x).$$

(A-1)

A.2.3.2   Standard Deviation

The variance or mean square error of a distribution provides one measurement of the degree to which the distribution is spread out. That is, the lack of "certainty" in the distribution's values. While the variance is useful in quantifying the spread of a distribution, it is not suited to quantifying the uncertainty in distribution values. The reason for this is that it is a measure of the *square* of this value.

The quantity that serves the purpose is the square root of the variance. This quantity is called the *standard deviation*. The standard deviation provides a measure of the spread of a distribution in units that are the same as the variable described by the distribution. Moreover, the standard deviation can be used to characterize the distribution in terms of its mathematical form. Quantities that characterize probability distributions are called *statistics*. As we will see presently, the standard deviation is an important statistic for uncertainty analysis.

A.2.3.3   Definition of Uncertainty

We have stated that the standard deviation of a distribution provides a measure of the "spread" of the distribution. Since a distribution is a relationship between the value of a variable and its probability of occurrence, we see that the more spread out the distribution is, the less likely we are of localizing the variable near its mean or nominal value. In other words, the uncertainty in the value of a variable is synonymous with the spread of its probability distribution. This leads to a statistical definition of measurement uncertainty:

$$u_x = \sigma_x = \sqrt{\text{var}(x)} \,.$$

(A-2)

Suppose we express a measured value for a variable x in terms of its true value μ_x and its measurement error ε_x

$$x = \mu_x + \varepsilon_x \,.$$

The variance in x, given a true value μ_x, is therefore given by

$$\text{var}(x) = \text{var}(\mu_x + \varepsilon_x) \,.$$

---

[30] The quantity $f(x)$ is the mathematical function that describes the form or shape of a probability distribution. It will be discussed at length later.

But there is no variance in the true value. For a given measurement, it is not a random variable, but is, instead, a fixed property of the measurand.[31] Consequently, we can write

$$\begin{aligned} \text{var}(x) &= \text{var}(\mu_x + \varepsilon_x) \\ &= \text{var}(\varepsilon_x). \end{aligned} \tag{A-3}$$

This expression comprises an important axiom in uncertainty analysis.

> **Axiom 2:** The variance in a measured variable is equal to the variance in the measurement error.

Given that the square root of the variance of a variable x is the standard deviation in x, and that the standard deviation in x is the uncertainty in x, we can use Axiom 2 to write

$$\begin{aligned} u_x &= \sigma_x \\ &= \sqrt{\text{var}(x)} \\ &= \sqrt{\text{var}(\varepsilon_x)}, \end{aligned} \tag{A-4}$$

which yields Axiom 3:

> **Axiom 3:** The uncertainty in a measurement is the standard deviation in the measurement error.

$$u_x = \sigma_{\varepsilon_x}. \tag{A-5}$$

Axiom 3 provides us with a statistical variable that can be used in decision analysis. Defined in this way, an uncertainty estimate is not just a number that we are required by governing standards to come up with but is, instead, a mathematical quantity that has a variety of uses, as we have alluded to and will discuss in more detail presently.

Note that the term "uncertainty" is sometimes used in reference to something called the "expanded uncertainty." This is an unfortunate use of the term in that the uncertainty and the expanded uncertainty differ in both magnitude and character. Considerable confusion and expense has come about due to the semantic similarity of the two terms. Expanded uncertainty will be discussed in detail later.

A.2.4 Combining Uncertainties

An important benefit of the above statistical definition of uncertainty is that it leads to a simple prescription for combining uncertainties that takes into account correlations between errors. This prescription emerges from a property of the variance called the variance addition rule.

---

[31] Note that, in a sample of measurements, the true value may vary randomly from measurement to measurement. The standard deviation due to the random error in the true value and the random error in the measurement system are included in the standard deviation of the sample, which comprises an estimate of the measurement process *repeatability*.

A.2.4.1   Variance Addition Rule

Suppose that a variable x is a linear function of two measured variables x_1 and x_2

$$x = a_1 x_1 + a_2 x_2 .$$

The variance in x is given by

$$\begin{aligned}
\mathrm{var}(x) &= \mathrm{var}(a_1 x_1 + a_2 x_2) \\
&= a_1^2 \, \mathrm{var}(x_1) + a_2^2 \, \mathrm{var}(x_2) + 2 a_1 a_2 \, \mathrm{cov}(x_1, x_2),
\end{aligned} \tag{A-6}$$

where the $\mathrm{cov}(x_1,x_2)$ term is the called the "covariance" between x_1 and x_2.

In reference to the combination of measurement errors, we express each of x_1 and x_2 as a true value plus error

$$\begin{aligned}
x &= a_1 x_1 + a_2 x_2 \\
&= a_1(\mu_{x1} + \varepsilon_{x1}) + a_2(\mu_{x2} + \varepsilon_{x2}) . \\
&= \mu_x + a_1 \varepsilon_{x1} + a_2 \varepsilon_{x2} .
\end{aligned} \tag{A-7}$$

The variance in x is then written

$$\begin{aligned}
\sigma_x^2 &= \mathrm{var}(a_1 \varepsilon_{x1} + a_2 \varepsilon_{x2}) \\
&= a_1^2 \, \mathrm{var}(\varepsilon_{x1}) + a_2^2 \, \mathrm{var}(\varepsilon_{x2}) + 2 a_1 a_2 \, \mathrm{cov}(\varepsilon_{x1}, \varepsilon_{x2}) \\
&= a_1^2 u_{\varepsilon_1}^2 + a_2^2 u_{\varepsilon_2}^2 + 2 a_1 a_2 \, \mathrm{cov}(\varepsilon_{x1}, \varepsilon_{x2}),
\end{aligned} \tag{A-8}$$

where the quantity $\mathrm{cov}(\varepsilon_{x1}, \varepsilon_{x2})$ is the covariance between the errors ε_{x1} and ε_{x2}.

A.2.4.2   The Correlation Coefficient

The covariance between two variables can be expressed in terms of a correlation coefficient defined according to

$$\rho(\varepsilon_{x1}, \varepsilon_{x2}) = \frac{\mathrm{cov}(\varepsilon_{x1}, \varepsilon_{x2})}{u_{x1} u_{x2}}, \tag{A-9}$$

which yields the expression

$$\sigma_x^2 = a_1^2 u_{x1}^2 + a_2^2 u_{x2}^2 + 2 a_1 a_2 \rho(\varepsilon_{x1}, \varepsilon_{x2}) u_{x1} u_{x2} . \tag{A-10}$$

A.2.4.3   The Combined Uncertainty

Since the uncertainty in measurement is the standard deviation of the measurement error and the standard deviation is the square root of the variance, the combined uncertainty in the measurement of x_1 and x_2 is given by

$$u_x = \text{Uncertainty}(x)$$
$$= \text{Uncertainty}(a_1 x_1 + a_2 x_2)$$
$$= \sigma_x$$
$$= \sqrt{a_1^2 u_{\varepsilon_{x1}}^2 + a_2^2 u_{\varepsilon_{x2}}^2 + 2 a_1 a_2 \rho(\varepsilon_{x1}, \varepsilon_{x2}) u_{\varepsilon_{x1}} u_{\varepsilon_{x2}}}.$$

(A-11)

An important characteristic of this result is that it is <u>not</u> simply the RSS value of the uncertainties in x_1 and x_2. The correlation term, that is absent from simple RSS combinations, may be significant in certain situations. In fact, if ε_{x1} and ε_{x2} are positive linear functions of one another, the correlation coefficient is equal to $+1$, and the combined uncertainty is

$$u_x = \sqrt{a_1^2 u_{\varepsilon_{x1}}^2 + a_2^2 u_{\varepsilon_{x2}}^2 + 2 a_1 a_2 u_{\varepsilon_{x1}} u_{\varepsilon_{x2}}}$$
$$= \sqrt{(a_1 u_{\varepsilon_{x1}} + a_2 u_{\varepsilon_{x2}})^2}$$
$$= a_1 u_{\varepsilon_{x1}} + a_2 u_{\varepsilon_{x2}}.$$

(A-12)

Conversely, if ε_{x1} and ε_{x2} are not related in any way, then $\rho(\varepsilon_{x1}, \varepsilon_{x2}) = 0$, and the uncertainty in x is given by

$$u_x = \sqrt{a_1^2 u_{\varepsilon_{x1}}^2 + a_2^2 u_{\varepsilon_{x2}}^2},$$

(A-13)

which is just the RSS combination of uncertainties for x_1 and x_2. Variables for which $\rho(\varepsilon_{x1}, \varepsilon_{x2}) = 0$ are said to be **statistically independent**. Many measurement errors fall into this category.

A.2.4.4 Degrees of Freedom

The degrees of freedom for a combined uncertainty u, given by

$$u = \sqrt{\sum_{i=1}^{k} a_i u_i^2},$$

(A-14)

can be computed using the Welch-Satterthwaite relation

$$\nu = \frac{u^4}{\sum_{i=1}^{k} \dfrac{a_i^4 u_i^4}{\nu_i}}.$$

(A-15)

The degrees of freedom for an uncertainty estimate is a measure of the amount of information that was employed in making the estimate.

A.3 Estimating Uncertainty

Up to this point, we have seen how errors are composed of error components, each of which is the sum of errors arising from sources of error encountered in the measurement process. We now turn our attention to the steps involved in computing estimates for the uncertainty in the source errors.

A.3.1   The Nature of Uncertainty Estimates

By their nature, uncertainty analyses are <u>approximate</u>. Uncertainty analysis requires us to make inferences about populations of statistical variables based on sampled data or on recollected experience. We rarely have all the information we need to develop precise statements of uncertainty.

Despite this, approximate uncertainty estimates can still be extremely useful quantities for decision making in that errors in uncertainty estimates tend to be small relative to the magnitudes of measured quantities. This is embodied in the statement that "errors in estimating uncertainty are 2nd order." To put this in perspective, we say that measured values are "zeroth" order, errors in these values are "first order," and errors in estimating the uncertainty of the measurement errors are second order. If measurement errors are small, then the uncertainty in the errors tends to be smaller still.

This does not mean we can be sloppy in estimating uncertainty. The better the estimate, the more valid our decisions based on measurements. The general recommendation is to attempt to make the <u>best</u> estimates possible. Note that this does not mean making the most conservative estimates possible. For technology management, seeking a conservative uncertainty estimate makes about as much sense as going to the hardware store and asking for a "conservative" tape measure. If conservatism is desired, the way to do enforce it is not with fudged uncertainty estimates but, rather, by specifying high levels of confidence in computing confidence limits or "expanded uncertainties."   This subject will be discussed later.

A.3.2   Computing Methods
A.3.2.1   Data Sampling

One way to estimate the uncertainty in an error source is to analyze a sample of measurements. The analysis consists primarily of computing a sample mean and standard deviation.

Taking Samples

In taking samples of measurements, we collect the results of some number of measurements. In collecting these results, we ensure that each measurement is both **independent** and **representative**. Measurements are independent if, in measuring one value, we do not affect or influence the selection of the measurement of another. Measurements are representative if their values are typical of the variable of interest.

Computing Statistics

Once a sample of measurements is taken, we estimate the characteristics of the population the sample was drawn from. In this, we make inferences about the population from certain statistics of the sampled data, i.e., from the sample mean and standard deviation.

A.3.2.2   Heuristic Methods

Another way to estimate the uncertainty in an error source is to employ heuristic methods. A heuristic estimate is an estimate made in the absence of recorded sampled data. Such an estimate is based on engineering judgment or on recollected experience. In making a heuristic estimate, we follow a simple procedure:

- Define error limits
- Estimate containment probabilities
- Estimate degrees of freedom.

A.3.3   Categories of Estimates

The manner in which we attempt to quantify the probability distributions of measurement errors falls into two broad categories labeled Type A (statistical) and Type B (heuristic).[32]

A.3.3.1   Type A Estimation

In Type A estimation, we attempt to infer various statistics of the population from data sampled from the population. For uncertainty analysis, the relevant statistics are the **mean** and **standard deviation** of the population. We approximate these statistics with the **sample mean** and **sample standard deviation**, respectively. For these approximations, we need to take into account the **degrees of freedom** of the sample from which they are computed.

The Sample Mean

A **sample mean** $\bar{x}$ is computed in a straightforward manner from a sample of measurements x_1, $x_2, \cdots, x_n$, where n is the sample size.

$$\bar{x} = \frac{1}{n}\sum_{i=1}^{n} x_i . \tag{A-16}$$

The sample mean is used as an estimate of the value that we expect to get when we make a measurement. This "expectation value" is called the **population mean**. A sample mean is called a **robust estimate**, if it approaches the population mean as the sample size increases.

The Sample Standard Deviation

In addition to inferring the expectation or mean value of a population, a measurement sample can also be used to estimate how much the population is spread about this value. As discussed earlier, the variable that quantifies this spread is called the **sample standard deviation**. The sample standard deviation is considered a robust estimate if it approaches the population standard deviation as the sample size increases.

For a given error source, we approximate the uncertainty due to error from the source by setting it equal to the sample standard deviation. Thus, for the measured variable x, the Type A uncertainty due to random error (repeatability), for example, is expressed as

$$u_{x,ran} = s_{x,ran}$$

$$= \sqrt{\frac{1}{n_x - 1}\sum_{j=1}^{n_x}(x_j - \bar{x})^2} , \tag{A-17}$$

---

[32] Note that this categorization applies to the manner in which estimates are made, not to the types of errors encountered. In the past, error types were usually classified as either random or systematic. These classifications are not related to Type A or B analysis designations.

where, the index j ranges over all sampled values of x. The sample mean $\bar{x}$ is defined as before. The number $\boldsymbol{n_x}$ is the **sample size** for the error source x. The above expression computes the **random uncertainty** in the measurement of x. As discussed earlier, this uncertainty arises from errors in measurement that are random with respect to magnitude and direction over the course of taking the measurement sample.

This uncertainty represents the random uncertainty in making a <u>single</u> measurement. Of equal importance is the **uncertainty in the mean value** obtained with a given measurement process. This uncertainty is obtained from something called the **sampling distribution** and is written

$$u_{\bar{x},ran} = \frac{s_{x,ran}}{\sqrt{n_x}}.$$

(A-18)

The Sample Degrees of Freedom

As Eq. (A-17) shows, the sample standard deviation is obtained by dividing the sum of the square of sampled deviations from the mean by the sample size minus one. As it turns out, dividing by n -1 instead of n is done to ensure that the standard deviation estimate is robust. The number n - 1 is called the **degrees of freedom** for the estimate.

The degrees of freedom for an estimate is the number of independent pieces of information that go into computing the estimate. The greater the degrees of freedom, the closer a sample estimate will be to its population counterpart. Because of this, the degrees of freedom is a useful quantity for establishing confidence limits and other decision variables. We will discuss degrees of freedom in more general terms later when we examine combinations of uncertainties.

Confidence Limits

The statistics obtained from a measurement sample can be used to compute limits that bound the mean value of the measurand with a specified level of confidence. The procedure utilizes the sample mean, standard deviation and degrees of freedom. If the variable of interest is normally distributed, confidence limits of

$$\bar{x} \pm t_{\alpha/2,v} s_{\bar{x}}$$

(A-19)

are said to contain the population mean value with $(1 - \alpha) \times 100\%$ confidence. In the above expression, the variable $t_{\alpha/2,v}$ is the t-statistic for a two-sided confidence level of $1 - \alpha$ and v degrees of freedom.

Note that the expression for computing confidence limits differs from the above if the measured variable is not normally distributed. For example, if the uniform distribution applies, the confidence limits would be computed according to

$$\bar{x} \pm P\sqrt{3} s_{\bar{x}},$$

where $P = \%$ confidence / 100.

A.3.3.2   Type B Estimation

A Type B estimate is obtained by drawing on recollected experience concerning the values of measured quantities or on knowledge of the errors in these quantities is. Type B estimates are made in the absence of recorded samples of measurement data. This does not mean that a Type B estimate is, by nature, obtained haphazardly. The process by which such estimates are arrived at can be cast in a structured format.

Heuristic Estimation Process

In making Type B estimates, we are free to draw on experience in any way that leads to consistent and reliable uncertainty estimates. A structured approach that has proved to be fruitful involves the following steps:

- Estimate error containment limits
- Estimate a containment probability
- Estimate the degrees of freedom
- Assume an underlying probability distribution and calculate the standard deviation

Estimate Containment Limits

Containment limits are limits that are said to "contain" or "bound" values of a variable of interest. These limits can contain either measured values or measurement errors. In developing containment limits, it is recommended that the best available experience with respect to containment limits be sought.

Containment limits are usually attribute tolerances, SPC control limits or other limits estimated from experience. An example of such limits would read something like "about 90% of the 50 or so cases observed, have fallen within ±10 psi." In this statement, the containment limits are ±10 psi, and the 90% figure represents a **containment probability**.

Estimate a Containment Probability

In statistics, we often use a sampled mean and standard deviation to estimate limits that can be said to bound deviations from the mean of a population. We call these limits **confidence limits**. The probability that the confidence limits will bound deviations from the mean is called the **confidence level**.

When making heuristic estimates of uncertainty, we instead use the terms **containment limits** and **containment probability**.

The containment probability is the probability that a set of containment limits will bound measured values or errors. It is recommended that the best available information be used in estimating containment probabilities. For example, if calibration history is available for a toleranced attribute, the number of observed in-tolerances divided by the number of calibrations would be a good estimate of the containment probability. The containment limits would, of course, be the tolerance limits.

Determine Degrees of Freedom

A value for the degrees of freedom can be found for a Type B estimate of uncertainty just as for an uncertainty or standard deviation estimate obtained from a random sample.

If a Type B estimate is obtained solely from containment limits and a containment probability, then the degrees of freedom is usually taken to be infinity. If an uncertainty estimate was an end in itself, this practice would not cause any difficulties. However, *uncertainty estimates have no intrinsic value*. They are merely statistics that are used to make inferences, establish error containment limits (e.g., attribute tolerance limits), compute risks. etc. More will be said on this later under the topic Type B Formats.

Compute the Standard Deviation

The steps in the process of obtaining Type B estimates of uncertainty are as follows:

1. Estimate the Expected Value for the Population.

    The expected value is usually a design value or "mode" value for the variable of interest. For example, the expected value for a 12 VDC source is 12 volts. The expected value for the resolution error is zero.

2. Estimate the Population Spread and the Degrees of Freedom. The population spread is the standard deviation of the underlying distribution for errors in the variable of interest. For normally distributed variables, the standard deviation is estimated according to

$$\text{Standard Deviation} = \frac{L}{t_{\alpha,\nu}}, \tag{A-20}$$

    where L is a containment limit, and ν is the degrees of freedom. The variable α is related to the containment probability p according to

$$
\begin{aligned}
\alpha &= (1-p)/2, \quad &\text{for two-sided containment limits} \\
\alpha &= 1-p, \quad &\text{for a one-sided containment limit}.
\end{aligned}
\tag{A-21}
$$

Example

To illustrate the Type B estimation process, we consider the uncertainty due to the bias of a reference attribute. In this illustration, imagine that we use a tape measure to measure the distance traveled by a car and a stopwatch to measure the elapsed time. The purpose of these measurements is to obtain the speed of the car.

Suppose that in an interval of time t we measure the distance traveled by the car to be X meters. We can make repeated measurements for the same time interval and use the data to estimate the random uncertainty in the measurement process, as discussed above. But we suspect that there are other uncertainties to account for. For one thing, we suppose that there is likely to be some systematic error present in each measurement due to a possible bias in the tape measure (we will ignore errors in the time measurement to simplify the discussion). This bias arises primarily from errors in the tape measure's manufacturing process and to stresses experienced during use.

We estimate the tape measure's bias uncertainty as follows. Apart from thermal expansion or contraction, response to stress or secondary effects, we know that the bias of the tape measure we use is a fixed quantity that persists from measurement to measurement. However, we don't know what the bias is. All we know is that the tape measure was drawn at random from its population and that each member of the population has its own bias. If we were to somehow employ a perfect measuring device and measure the bias of each member of the population, we would see that these measurements would follow some kind of probability distribution. The spread or standard deviation in this distribution is the uncertainty in the bias of a tape measure drawn from the population.

To estimate the tape measure bias uncertainty, we use the manufacturer's specifications, together with an estimated confidence level that the tape measure is within specification. Suppose that the tolerance limits for the tape measure are ± 0.5mm and that the confidence that the tape measure is in-tolerance is 95%. We compute the standard deviation for the tape measure bias according to

$$\sigma_{bias} = \frac{0.5 \text{ mm}}{t_{0.025,\nu}},$$

where the variable $t_{0.025,\nu}$ is a coverage factor representing an in-tolerance confidence of 95% and ν degrees of freedom. By comparing this expression with what we discussed earlier, we see that the tolerance limits serve as the error **containment limits**, and that the confidence level suffices as the **containment probability**.

We now estimate ν and $t_{0.025,\nu}$. To do this, we need to make certain assumptions about the probability distribution for tape measure biases. Given what we know about the tape measure's distribution, which is very little, we use the most appropriate distribution we can think of for the problem in front of us. This distribution is the **normal distribution**. As for the degrees of freedom, we would need some information about the confidence level used or about the reliability of the ± 0.5mm tolerance limits. We don't have this information, so we will assume infinite degrees of freedom, i.e., we will set $\nu = \infty$. From statistical tables, we obtain a value $t_{0.025,\infty} = 1.960$, which yields

$$\sigma_{bias} = \frac{0.5 \text{ mm}}{1.96} = 0.255 \text{ mm}.$$

This is our estimate for the uncertainty in the bias of the tape measure. Accordingly, we write

$$u_{bias} = 0.255 \text{ mm}.$$

Heuristic uncertainty estimates for other error sources are conducted in a similar manner. As a final note, we mention that a different result would have been obtained if we had knowledge about the degrees of freedom associated with the error containment limits (tolerance limits) and the containment probability (confidence level). If the degrees of freedom were known, we would have used the Student's t distribution rather than the normal distribution. Developing heuristic estimates with nonzero degrees of freedom and with other error distributions is discussed in the next section.

Type B Formats

For Type B estimates to be combined with Type A estimates in a meaningful way, the degrees of freedom associated with both must be determined. This ensures that the degrees of freedom for the combined uncertainty will be statistically valid — a necessity for computing confidence limits and for other possible applications of the combined uncertainty, such as the analysis of decision risks.

The process of assembling the recollected experience and other technical data needed for developing Type B degrees of freedom estimates can be facilitated through the use of standardized formats. These formats also provide a means of computing Type B estimates and placing them on a statistical footing.

A.4   Multivariate Uncertainty Analysis

Frequently, the value of a quantity of interest is obtained by measuring the values of constituent quantities. An example is the measurement of velocity, obtained through measurements of time and distance. In such cases, we are required to build an **error model** to work from in developing an expression for the uncertainty in the quantity of interest.

A.4.1   Error Modeling

Error modeling consists of identifying the various components of error and establishing their relationship to one another. The guiding expression for this process is the **system equation**.

A.4.1.1   The System Equation

The system equation is the expression for the variable being sought in terms of its measurable components. Establishing the system equation is often the most difficult part of the process. If the system equation can be determined, then uncertainty analysis becomes almost automatic.

For purposes of illustration, we consider a two-component system variable. The expressions that ensue can easily be extended to system equations with arbitrary numbers of components.

Let the component variables of the system equation be labeled x and y. Then, if the variable of interest, labeled z, is expressed as a function of x and y, we have

$$z = z(x, y). \tag{A-22}$$

A.4.1.2   Error Components

Each measurable variable in the system equation is an error component. The contribution of each error component to the error in the variable z is obtained in a Taylor series expansion of Eq. (A-22)

$$\varepsilon_z = \left(\frac{\partial z}{\partial x}\right)\varepsilon_x + \left(\frac{\partial z}{\partial y}\right)\varepsilon_y + O(2), \tag{A-23}$$

where $O(2)$ indicates terms to second and higher order in the error variables. These terms are usually negligible and are dropped from the expression to yield

$$\varepsilon_z \cong \left(\frac{\partial z}{\partial x}\right)\varepsilon_x + \left(\frac{\partial z}{\partial y}\right)\varepsilon_y. \tag{A-24}$$

A.4.2   Computing System Uncertainty

Using Axiom 2 and the variance addition rule with this equation gives

$$u_z = \sqrt{a_x^2 u_x^2 + a_y^2 u_y^2 + 2a_x a_y \rho(\varepsilon_x, \varepsilon_y) u_x u_y} \,, \tag{A-25}$$

where the coefficients a_x and a_y are

$$a_x = \left(\frac{\partial z}{\partial x}\right), \qquad a_y = \left(\frac{\partial z}{\partial y}\right), \tag{A-26}$$

and the uncertainties are

$$u_x = \sqrt{\mathrm{var}(\varepsilon_x)}, \qquad u_y = \sqrt{\mathrm{var}(\varepsilon_y)} \,. \tag{A-27}$$

A.4.2.1 Process Uncertainties

Various sources of error contribute to each error component of a measurement. Most commonly encountered are the measurement process errors described earlier

- Measurement Bias
- Random Error
- Resolution Error
- Digital Sampling Error
- Computation Error
- Operator Bias
- Stress Response Error
- Environmental/Ancillary Factors Error

Labeling each of the relevant error sources with a number designator, we can write the error in a component x as

$$\varepsilon_x = \varepsilon_{x1} + \varepsilon_{x2} + \mathrm{L}\ \varepsilon_{xn} \,, \tag{A-28}$$

and the uncertainty in the measurement of x becomes

$$\begin{aligned}
u_x^2 = u_{x1}^2 + u_{x2}^2 + \mathrm{L} + u_{xn}^2 + 2\rho(\varepsilon_{x1}, \varepsilon_{x2}) u_x u_{x2} \\
+ 2\rho(\varepsilon_{x1}, \varepsilon_{x3}) u_{x1} u_{x3} + \mathrm{L} + 2\rho(\varepsilon_{xn-1}, \varepsilon_{xn}) u_{xn-1} u_{xn} \,.
\end{aligned} \tag{A-29}$$

A.4.2.2 Cross-Correlations

The final topic of this section is the cross-correlations between error sources of different error components. Cross-correlations occur when different components of a variable are measured using the same device, in the same environment, by the same operator or in some other way that would lead us to suspect that the measurement errors in the two components might be correlated.

If the cross-correlation between the *ith* and the *jth* process errors of the measured variables x and y is denoted by $\rho(\varepsilon_{xi}, \varepsilon_{yj})$, then the correlation coefficient between x and y is given by

$$\rho_{xy} = \frac{1}{u_x u_y} \sum_{i=1}^{n_x} \sum_{j=1}^{n_y} \rho(\varepsilon_{xi}, \varepsilon_{yj}) u_x u_{yj} \,, \tag{A-30}$$

where n_x and n_y are the number of process errors for the x and y error components, respectively, and the component uncertainties are

$$u_x^2 = \sum_{i=1}^{n_x} u_{xi}^2 + 2 \sum_{i=1}^{n_x-1} \sum_{j=i+1}^{n_x} \rho(\varepsilon_{xi}, \varepsilon_{xj}) u_{xi} u_{xj} , \qquad (A\text{-}31)$$

and

$$u_y^2 = \sum_{i=1}^{n_y} u_{yi}^2 + 2 \sum_{i=1}^{n_y-1} \sum_{j=i+1}^{n_y} \rho(\varepsilon_{yi}, \varepsilon_{yj}) u_{yi} u_{yj} . \qquad (A\text{-}32)$$

A.5   Expanded Uncertainty

The expanded uncertainty is a limit obtained by multiplying an uncertainty estimate by a specified **coverage factor**. For technology management purposes, it is desirable to relate the coverage factor to a desired confidence level. If this is done, then the expanded uncertainty serves as a **confidence limit** that can be said to bound errors with a stated degree of confidence.

Suppose, for example, that a mean value $\bar{x}$ and an uncertainty estimate u have been obtained for a variable x, along with a degrees of freedom estimate ν. Confidence limits $\pm U_{p,\nu}$ that bound errors in x with some probability p can be computed as[33]

$$U_{p,\nu} = t_{\alpha,\nu} u , \qquad (A\text{-}33)$$

where the coverage factor $t_{\alpha,\nu}$ is the familiar Student's t-statistic with $\alpha = (1 - p) / 2$ for two-sided limits and $\alpha = (1 - p)$ for single-sided limits. In using the expanded uncertainty to indicate our confidence in the estimate $\bar{x}$, we would say that the value of x is estimated as $\bar{x} \pm U_{p,\nu}$ with $p \times 100\%$ confidence.

Some individuals and agencies employ a fixed, arbitrary number to use as a coverage factor. The argument for this practice is primarily based on the assertion that the degrees of freedom for Type B estimates cannot be rigorously established. Accordingly, it makes little sense to attempt to determine a t-statistic based on a confidence level and degrees of freedom for a combined Type A/B uncertainty. Until recently, this assertion had been true. Methods now exist, however, that permit the determination of Type B degrees of freedom (See Appendix K). Given this, the practice of using a fixed number for all expanded uncertainty estimates is not recommended. Such estimates are, at best, only loosely related to confidence limits.

A.6   Test and Calibration Scenarios

This section discusses information obtained from measurements made during calibration with a focus on developing uncertainty estimates that are applicable to measurement decision risk analysis. Four calibration scenarios are discussed:

1. The measurement reference (MTE) measures the value of an attribute of the unit under test (UUT) that provides an output or stimulus.

2. The UUT measures the value of a reference attribute of the MTE that provides an output or stimulus.

---

[33] The distribution for the population of errors in x is assumed to be normal in this example; hence the use of the t-statistic for computing confidence limits.

3. The UUT and MTE each provide an output or stimulus for comparison using a bias cancellation comparator.

4. The UUT and MTE both measure the value of an attribute of a common device or artifact that provides an output or stimulus.

The information obtained includes an observed value, referred to as a "measurement result" or "calibration result," and an estimated uncertainty in the measurement error. For each scenario, a measurement equation is given that is applicable to the manner in which calibrations are performed and calibration results are recorded or interpreted.

The measurement scenarios turn out to be simple and intuitive. In each, the measurement result and the measurement error are separable, allowing the estimation of measurement uncertainty. For the purposes of this Handbook Annex, it is assumed in each scenario that the measurement result is an estimate of the value of the bias of the UUT attribute.

A.6.1 Basic Notation

The subscripts and variables designators in this Appendix are summarized in Table A-2. With this notation, measurement error is represented by ε_m, the error in a calibration result by ε_{cal} and the bias in the UUT attribute by $e_{UUT,b}$.

Table A-2. Basic Notation.

Notation	Description
e	an individual measurement process error, such as repeatability, resolution error, etc.
ε	combined errors comprised of individual measurement process errors
m	measurement
b	bias
cal	calibration
$true$	true value
n	nominal value

As stated in the introduction, specific measurement equations will be given for each calibration scenario. In each equation, quantities relating to the UUT are indicated with the notation x and quantities relating to the MTE with the notation y. For example, in Scenario 1, where the MTE directly measures the value of the UUT attribute, the relevant measurement equation is

$$y = x_{true} + \varepsilon_m , \qquad\qquad (A\text{-}34)$$

where y represents a measurement taken with the MTE, given a UUT attribute true value x_{true}, and ε_m is the measurement error. Variations of Eq. (A-34) will be encountered throughout this Appendix.[34]

---

[34] In taking a sample of values of y in a test or calibration, the quantity x_{true} may vary from measurement to measurement. Since contributions to e_{rep} due to these variations are not distinguishable from a random error

A.6.2 Measurement Uncertainties

Measurement errors and attribute biases are random variables that follow probability distributions. Each distribution is a relationship between the value of an error and its probability of occurrence. Distributions for errors that are tightly constrained correspond to low uncertainty, while distributions for errors that are widely spread correspond to high uncertainties. Mathematically, the uncertainty due to a particular error is equated to the spread in its distribution. This spread is just the distribution standard deviation which is defined as the square root of the distribution variance [A-1, A-2]. Letting u represent uncertainty and "var(ε)" the statistical variance of the distribution of an error ε, we write

$$u = \sqrt{\text{var}(\varepsilon)} \ .$$

(A-35)

This expression will be used in this Appendix as a template for estimating measurement process uncertainties encountered in the various calibration scenarios.

A.6.3 Measurement Error Sources

Typically, calibration scenarios feature the following set of measurement process errors or "error sources."

$e_{MTE,b}$ = bias in the measurement reference

e_{rep} = repeatability or "random" error

e_{res} = resolution error

e_{op} = operator bias

e_{other} = other measurement error, such as that due to environmental corrections, ancillary equipment variations, response to adjustments, etc.

A.6.3.1 Measurement Reference Bias

The error in a measurement reference attribute, at any instant in time, is composed of a systematic component and a random component. The systematic component is called "attribute bias." Attribute bias is an error component that persists from measurement to measurement during a "measurement session." Attribute bias excludes resolution error, random error, operator bias and other sources of error that are not properties of the attribute.[35]

A.6.3.2 Repeatability

Repeatability is a random error that manifests itself as differences in measured value from measurement to measurement during a measurement session. It should be said that random variations in UUT attribute value and random variations due to other causes are not separable

---

contribution to ε_m, the quantity e_{rep} is referred to as a "measurement process error," i.e., one emerging from the measurement process.

[35] For purposes of discussion, a measurement session is considered to be an activity in which a measurement or sample of measurements is taken under fixed conditions, usually for a period of time measured in seconds, minutes or, at most, hours.

from random variations in the value of the MTE reference attribute or any other error source. Consequently, whether e_{rep} manifests itself in a sample of measurements made by the MTE or by the UUT, it must be taken to represent a "measurement process error" rather than an error attributable to any specific influence.

A.6.3.3 Resolution Error

Reference attributes and/or UUT attributes may provide indications of sensed or stimulated values with some finite precision.

For example, a voltmeter may indicate values to four, five, six, etc., significant digits. A tape measure may provide length indications in meters, centimeters and millimeters. A scale may indicate weight in terms of kg, g, mg, etc. The smallest discernible value indicated in a measurement comprises the resolution of the measurement.

The basic error model for resolution error is

$$x_{indicated} = x_{sensed} + e_{res},$$

where x_{sensed} is a "measured" value detected by a sensor or provided by a stimulus, $x_{indicated}$ is the indicated representation of x_{sensed} and e_{res} is the resolution error.

A.6.3.4 Operator Bias

Because of the potential for operators to acquire measurement information from an individual perspective or to produce a systematic bias in a measurement result, it sometimes happens that two operators observing the same measurement result will systematically perceive or produce different measured values. The systematic error in measurement due to the operator's perspective or other tendency is referred to as **Operator Bias**.

Operator bias is a "quasi-systematic" error, the error source being the perception of a human operator. While variations in human behavior and response lend this error source a somewhat random character, there may be tendencies and predilections inherent in a given operator that persist from measurement to measurement.

The random contribution is included in the random error source discussed earlier. The systematic contribution is the operator bias.

A.6.3.5 Repeatability and Resolution Error

It is sometimes argued that repeatability is a manifestation of resolution error. To address this point, imagine three cases. In the first case, values obtained in a random sample of measurements take on just two values and the difference between them is equal to the smallest increment of resolution. If this is the case, we can conclude that "background noise" random variations are occurring that are beyond the resolution of the measurement. If so, we cannot include repeatability as an error source but must acknowledge that the apparent random variations are due to resolution error. Accordingly, the uncertainty due to resolution error should be included in the total measurement uncertainty but the uncertainty due to repeatability should not.

In the second case, values obtained in a random sample are seen to vary in magnitude substantially greater than the smallest increment of resolution. In this case, repeatability cannot be ignored as an error source. In addition, since each sampled value is subject to resolution error, resolution error should also be separately accounted for. Accordingly, the total measurement uncertainty must include contributions from both repeatability uncertainty and resolution uncertainty.

The third case is not so easily dealt with. In this case, values obtained in a random sample of measurements are seen to vary in magnitude somewhat greater than the smallest increment of resolution but not substantially greater. We perceive an error due to repeatability that is separable from resolution error but is partly due to it. It then becomes a matter of opinion as to whether to include repeatability and resolution error in the total measurement error. Until a clear solution to the problem is found, it is the opinion of the authors that both should be included in this case.

A.6.3.6   Other Error

Other measurement error is a catch-all label applied to errors such as those due to environmental corrections, ancillary equipment variations, response to adjustments, etc. For example, suppose that "other" error is due to an environmental factor, such as temperature, vibration, humidity or stray emf. In many cases, as in accommodating thermal expansion, the effect of an environmental factor can be corrected for. Such corrections usually rely on a measurement of the driving environmental factor.

When this happens, the attribute that measures the environmental factor is referred to as an ancillary attribute. An example would be a thermometer reading used to correct for thermal expansion in the measurement reference and the UUT attributes. Since an ancillary attribute is subject to error, as is any other attribute, this error can lead to an error in the environmental correction. The uncertainty in the error of the correction is a function of the uncertainty in the error due to the environmental factor.

For a more complete discussion on uncertainties due to environmental and other ancillary factors, see Ref [A-1].

A.6.4   Calibration Error and Measurement Error

For the scenarios discussed in this Appendix, the result of a calibration is taken to be the estimation of the bias $e_{UUT,b}$ of the UUT attribute. The error in the calibration result is represented by the quantity ε_{cal}. In all scenarios, the uncertainty in the estimation of $e_{UUT,b}$ is computed as the uncertainty in ε_{cal}. For some scenarios, ε_{cal} is synonymous with the measurement error ε_m or its negative $-\varepsilon_m$. In other scenarios, as in Scenario 2 where the UUT measures the MTE attribute, ε_m includes $e_{UUT,b}$. Since $e_{UUT,b}$ cannot be included in ε_{cal}, the latter of which is the error in the estimation of $e_{UUT,b}$, we have a situation where ε_{cal} and ε_m may not be of the same sign or magnitude.

A.6.5　UUT Attribute Bias

For calibrations, it is tacitly assumed that the UUT attribute of interest is assigned some design or "nominal" value x_n. The difference between the UUT attribute's true value, x_{true}, and the nominal value x_n is the UUT attribute's bias $e_{UUT,b}$. Accordingly, we can write

$$x_{true} = x_n + e_{UUT,b}.^{36} \tag{A-36}$$

In some cases, the UUT attribute is a passive attribute, such as a gage block or weight, whose attribute of interest is a simple characteristic like length or mass. In other cases the UUT is an active device, such as a voltmeter or tape measure, whose attribute consists of a reading or other output, like voltage or measured length. In the former case, the concepts of true value and nominal value are straightforward. In the latter case, some comment is needed.

As stated earlier, we consider the result of a calibration to be an estimate of the quantity $e_{UUT,b}$. From Eq. (A-36), we can readily appreciate that, if we can assign the UUT a nominal value x_n, estimating x_{true} is equivalent to estimating $e_{UUT,b}$. Additionally, we acknowledge that $e_{UUT,b}$ is an "inherent" property of the UUT, independent of its resolution, repeatability or other characteristic dependent on its application or usage environment. Accordingly, if the UUT's nominal value consists of a measured reading or other actively displayed output, the UUT bias must be taken to be the difference between the true value of the quantity being measured and the value internally sensed by the UUT, with appropriate environmental or other adjustments applied to correct this value to reference (calibration) conditions.

For example, imagine that the UUT is a steel yardstick whose length is a random variable following a probability distribution with a standard deviation arising from variations in the manufacturing process. Imagine now that the UUT is used under specified nominal environmental conditions. While under these conditions, repeatability, resolution error, operator bias and other error sources may come into play, the bias of the yardstick is systematically present, regardless of whatever chance relationship may exist between the length of the measured object, the closest observed "tick mark," the temperature of the measuring environment, the perspective of the operator, and so on.

A.6.6　MTE Bias

The value of the reference attribute of the MTE, against which the value of the UUT attribute is compared, has an inherent deviation $e_{MTE,b}$ from its nominal value or a value stated in a calibration certification or other reference document. Letting y_{true} represent the true value of the MTE attribute and letting y_n represent the MTE attribute nominal or assumed value, we have

$$y_{true} = y_n + e_{MTE,b}.^{37} \tag{A-37}$$

In some cases, the MTE attribute is a passive attribute, such as a gage block or weight, whose attribute is a simple characteristic like length or mass. In other cases the MTE is an active

---

[36] Note that Eq. (A-36) is not a measurement model as defined in the basic measurement equation given in Section 2.2.2. Rather, it is a statement of the relationship between the UUT attribute's true value, its nominal value and its inherent bias. Given this, if we view the attribute's nominal value as a "measurement" of its true value, then the relationship between the "measurement" error and the attribute bias is $\varepsilon_m = -e_{UUT,b}$.

[37] See Footnote 37.

device, such as a voltmeter or tape measure, whose attribute consists of a reading or other output, like voltage or measured length. In either case, it is important to bear in mind that $e_{MTE,b}$ is an inherent property of the MTE, exclusive of other errors such as MTE resolution or the repeatability of the measurement process. It may vary with environmental deviations, but can usually be adjusted or corrected to some reference set of conditions. An illustration of such an adjustment is given below in Scenario 1.

A.7 Calibration Scenarios

The four calibration scenarios identified in this Appendix's introduction are described in detail in the following discussions. The descriptions are not offered to serve as recipes to be followed as dogma but are, instead, intended to provide guidelines for developing uncertainty estimates relevant to each scenario. Hopefully, the structure and content of each description will assist in developing whatever mathematical customization is needed for specific measurement situations.[38]

In each scenario, we have a measurement of $e_{UUT,b}$, denoted δ, and a calibration error ε_{cal}. The general expression is

$$\delta = e_{UUT,b} + \varepsilon_{cal}.$$

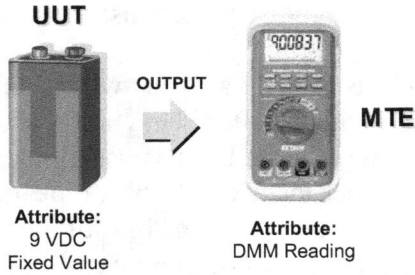

Figure A-8. Scenario 1

The MTE measures the value of a UUT Attribute. The output is the battery voltage.

Since $e_{UUT,b}$ is the quantity being estimated by calibration, as discussed earlier, the uncertainty of interest is understood to be the uncertainty in δ given the UUT bias $e_{UUT,b}$. Then, by Eq. (A-35), we have

$$u_{cal} = \sqrt{\mathrm{var}(\delta)}$$
$$\sqrt{\mathrm{var}(e_{UUT,b} + \varepsilon_{cal})} \tag{A-38}$$
$$= \sqrt{\mathrm{var}(\varepsilon_{cal})}.$$

---

[38] Examples demonstrating the procedures and concepts of this section are given in Chapter 6 of the Handbook.

A.7.1 Scenario 1: The MTE Measures the UUT Attribute Value

In this scenario, the UUT is a passive device whose calibrated attribute provides no reading or other metered output. Its output may consist of a generated value, as in the case of a voltage reference, or a fixed value, as in the case of a gage block.[39] The measurement equation is repeated from Eq. (A-34) as

$$y = x_{true} + \varepsilon_m , \qquad (A\text{-}39)$$

where y is the measurement result obtained with the MTE, x_{true} is the true value of the UUT attribute and ε_m is the measurement error.

The "measured" value provided by the UUT is its nominal value x_n, given in Eq. (A-36), so that

$$x_{true} = x_n + e_{UUT,b}. \qquad (A\text{-}40)$$

Substituting Eq. (A-40) in Eq. (A-39), we write the measurement equation as

$$y = x_n + e_{UUT,b} + \varepsilon_m .$$

The difference $y - x_n$ is a measurement of the UUT bias $e_{UUT,b}$. We denote this quantity by the variable δ and write

$$
\begin{aligned}
\delta &= y - x_n \\
&= e_{UUT,b} + \varepsilon_m \\
&= e_{UUT,b} + \varepsilon_{cal} .
\end{aligned}
\qquad (A\text{-}41)
$$

For this scenario, the calibration error ε_{cal} is equal to the measurement error ε_m and is comprised of MTE bias, measurement process repeatability, MTE resolution error, operator bias, etc. The appropriate expression is

$$\varepsilon_{cal} = e_{MTE,b} + e_{rep} + e_{res} + e_{op} + e_{other} . \qquad (A\text{-}42)$$

Since the UUT is a passive device in this scenario, resolution error, and operator bias arise exclusively from the use of the MTE, i.e., $e_{UUT,res}$ and $e_{UUT,op}$ are zero. In addition, the uncertainty due to repeatability is estimated from a random sample of measurements taken with the MTE. Still, variations in UUT attribute value may contribute to this estimate. However, random variations in UUT attribute value and random variations due to other causes are not separable from random variations due to the MTE. Consequently, as stated earlier, e_{rep} must be taken to represent a "measurement process error" rather than an error attributable to any specific influence. Given these considerations, the error sources e_{rep}, e_{res} and e_{op} in Eq. (A-42) are

$$
\begin{aligned}
e_{rep} &= e_{MTE,rep} \\
e_{res} &= e_{MTE,res} \\
e_{op} &= e_{MTE,op} ,
\end{aligned}
\qquad (A\text{-}43)
$$

---

[39] Cases where the MTE and UUT attributes each exhibit a displayed value are covered later as special instances of Scenario 4.

where $e_{MTE,rep}$ represents the repeatability of the measurement process. The "MTE" part of the subscript indicates that the uncertainty in the error will be estimated from a sample of measurements taken by the MTE.

In some cases, the error source e_{other} may need some additional thought. For example, suppose that e_{other} arises from corrections ensuing from environmental factors, such as thermal expansion. If measurements are made of the length of a UUT gage block using an MTE reference "super mike," it may be desired to correct measured values to those that would be attained at some reference temperature, such as 20 °C.

Let $\delta_{UUT,env}$ and $\delta_{MTE,env}$ represent thermal expansion corrections to the gage block and super mike dimensions, respectively. Then the mean value of the measurement sample would be corrected by an amount equal to[40]

$$\delta_{env} = \delta_{MTE,env} - \delta_{UUT,env}, \tag{A-44}$$

and the error in the corrections would be written

$$\begin{aligned} e_{other} &= e_{env} \\ &= e_{MTE,env} - e_{UUT,env}. \end{aligned} \tag{A-45}$$

From Eqs. (A-41) and (A-38), we can write the uncertainty in the calibration result δ as

$$u_{cal} = \sqrt{\text{var}(\varepsilon_{cal})}, \tag{A-46}$$

where, by Eq. (A-42),

$$\begin{aligned} \text{var}(\varepsilon_{cal}) &= \text{var}(e_{MTE,b}) + \text{var}(e_{rep}) + \text{var}(e_{res}) + \text{var}(e_{op}) + \text{var}(e_{other}) \\ &= u_{MTE,b}^2 + u_{rep}^2 + u_{res}^2 + u_{op}^2 + u_{other}^2, \end{aligned} \tag{A-47}$$

and

$$\begin{aligned} u_{rep} &= u_{MTE,rep} \\ u_{res} &= u_{MTE,res} \\ u_{op} &= u_{MTE,op}. \end{aligned} \tag{A-48}$$

For this scenario, no correlations are present between the error sources shown in Eq. (A-47). Hence the simple RSS uncertainty combination. This may not be true for correlations *within* some of the terms, as may be the case when $e_{other} = e_{env}$. In this case, we would have

$$u_{other} = \sqrt{u_{MTE,env}^2 + u_{UUT,env}^2 - 2\rho_{env} u_{MTE,env} u_{UUT,env}}. \tag{A-49}$$

If the same temperature measurement device (e.g., thermometer) is used to make both the UUT and MTE corrections, we would have $\rho_{env} = 1$, and

---

[40] The form of this expression arises from the fact that thermal expansion of the gage block results in an inflated gage block length, while thermal expansion of the supermike results in applying additional thimble adjustments to narrow the gap between the anvil and the spindle, resulting in a deflated measurement reading.

$$u_{other} = \sqrt{u_{MTE,env}^2 + u_{UUT,env}^2 - 2u_{MTE,env}u_{UUT,env}}$$
$$= \left| u_{MTE,env} - u_{UUT,env} \right|. \tag{A-50}$$

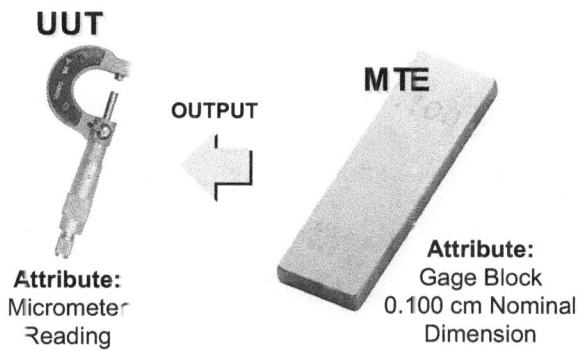

Figure A-9. Scenario 2

The UUT measures the value of an MTE attribute. The output is the gage block dimension.

A.7.2 Scenario 2: The UUT Measures the MTE Attribute Value

In this scenario, the MTE is a passive device whose reference attribute provides no reading or other metered output. Its output may consist of a generated value, as in the case of a voltage reference, or a fixed value, as in the case of a gage block.[41] The measurement equation is a variation of Eq. (A-34)

$$x = y_{true} + \varepsilon_m , \tag{A-51}$$

where x is the value measured by the UUT, y_{true} is the true value of the MTE attribute being measured and ε_m is the measurement error. Denoting the nominal or indicated value of the MTE as y_n, we can write

$$y_{true} = y_n + e_{MTE,b} , \tag{A-52}$$

where $e_{MTE,b}$ is defined in Eq. (A-37). Substituting Eq. (A-52) in Eq. (A-51) gives

$$x = y_n + e_{MTE,b} + \varepsilon_m , \tag{A-53}$$

and

$$\delta = e_{MTE,b} + \varepsilon_m \tag{A-54}$$

where δ is the measurement of the UUT bias, given by

$$\delta = x - y_n . \tag{A-55}$$

For this scenario, the measurement error is given by

$$\varepsilon_m = \varepsilon_{UUT,b} + e_{rep} + e_{res} + e_{op} + e_{other} , \tag{A-56}$$

where $e_{UUT,b}$ is the UUT bias defined in Eq. (A-36), e_{rep} is the repeatability of the measurement

---

[41] Cases where the MTE and UUT attributes each exhibit a displayed value are covered later as special instances of Scenario 4.

process as evidenced in the sample of measurements taken with the UUT, e_{res} is the resolution error of the UUT and e_{op} is operator bias associated with the use of the UUT

$$e_{rep} = e_{UUT,rep}$$
$$e_{res} = e_{UUT,res} \qquad \text{(A-57)}$$
$$e_{op} = e_{UUT,op}.$$

The error source e_{other} may need to include mixed contributions as described in Scenario 1.

Substituting Eq. (A-56) in Eq. (A-54) and rearranging gives

$$\delta = e_{UUT,b} + e_{MTE,b} + e_{rep} + e_{res} + e_{op} + e_{other} \qquad \text{(A-58)}$$

where e_{rep}, e_{res}, and e_{op} are defined in Eq. (A-57).

As before, we obtain an expression that is separable into a measurement δ of the UUT bias, $e_{UUT,b}$ and an error ε_{cal} given by

$$\varepsilon_{cal} = e_{MTE,b} + e_{rep} + e_{res} + e_{op} + e_{other}. \qquad \text{(A-59)}$$

By Eq. (A-38), the uncertainty in the $e_{UUT,b}$ estimate in Eq. (A-58) is

$$u_{cal} = \sqrt{\text{var}(\varepsilon_{cal})}, \qquad \text{(A-60)}$$

where

$$\text{var}(\varepsilon_{cal}) = u_{MTE,b}^2 + u_{rep}^2 + u_{res}^2 + u_{op}^2 + u_{other}^2. \qquad \text{(A-61)}$$

A.7.3   Scenario 3: MTE and UUT Output Comparison (Comparator Scenario)

In this scenario, a device is used to compare UUT and MTE values where both the UUT and the MTE provide an output value or stimulus. For this scenario, the device is called a "comparator." It is worthwhile to consider the following procedure:

1. The MTE is placed in the comparator.

2. The comparator indication or reading y is noted. This indication or reading is taken to correspond to the MTE nominal or reading value y_n.

3. The MTE is removed and the UUT is placed in the comparator.

4. The comparator indication or reading x is noted.

5. The difference δ is calculated, where

$$\delta = x - y \qquad \text{(A-62)}$$

comprises a measurement of the UUT bias $e_{UUT,b}$. The UUT corrected value, denoted x_c, is then given by

$$x_c = y_n + \delta. \qquad \text{(A-63)}$$

With this procedure, any bias introduced by the comparator is cancelled in Eq. (A-62).[42]

---

[42] In many comparator calibrations, the comparator device is made up of two measurement arms and a meter or other indicator. The UUT and the MTE are placed in different arms of the comparator and the difference between the values is displayed by the indicating device. In such cases, if the UUT and MTE swap locations, and the average of the differences is recorded, then bias cancellation is achieved as in the procedure described above. Of course, Eqs. (A-74) and (A-75) would need to be modified to accommodate any additional measurement process errors, such as additional contributions due to comparator resolution error.

If this swapping procedure is not followed, the comparator bias is not cancelled and the applicable scenario becomes a variation of Scenario 1 or 2 in which the comparator, taken in aggregate, is treated as the MTE, with the reference item, indicating device, comparator arms, etc. acting as components. The estimation of the bias uncertainty of the aggregate MTE is the subject of multivariate uncertainty analysis, described in Annex 3.

UUT

MTE

Attribute:
10 gm
Nominal Mass

Attribute:
10.000 gm
Nominal Mass

Figure A-10. Scenario 3

Measured values of the UUT and MTE attributes are compared using a comparator. The outputs are the weights of the masses.

In keeping with the basic notation, the indicated value y can be expressed as

$$y = y_{true} + \varepsilon_{MTE,m} \tag{A-64}$$

and the indicated value x can be written

$$x = x_{true} + \varepsilon_{UUT,m} \tag{A-65}$$

where $\varepsilon_{MTE,m}$ is the measurement error involved in the use of the comparator to measure the MTE attribute value and $\varepsilon_{UUT,m}$ is the measurement error involved in the use of the comparator to measure the UUT attribute value.

By Eq. (A-37), we can write

$$y_{true} = y_n + e_{MTE,b} \tag{A-66}$$

and

$$x_{true} = x_n + e_{UUT,b} \tag{A-67}$$

Substituting Eq. (A-66) in Eq. (A-64) gives

$$y = y_n + e_{MTE,b} + \varepsilon_{MTE,m} \tag{A-68}$$

and substituting Eq. (A-67) in Eq. (A-65) yields

$$x = x_n + e_{UUT,b} + \varepsilon_{UUT,m} . \tag{A-69}$$

Using Eqs. (A-68) and (A-69) in Eq. (A-62), we can write

$$\delta = x - y$$
$$= x_n - y_n + e_{UUT,b} - e_{MTE,b} + (\varepsilon_{UUT,m} - \varepsilon_{MTE,m}),$$

so that

$$e_{UUT,b} = \delta - (x_n - y_n) + e_{MTE,b} - (\varepsilon_{UUT,m} - \varepsilon_{MTE,m}) . \tag{A-70}$$

In most calibrations involving comparators $x_n = y_n$ and Eq. (A-70) becomes [43]

$$e_{UUT,b} = \delta + e_{MTE,b} - (\varepsilon_{UUT,m} - \varepsilon_{MTE,m}).$$

(A-71)

Then, as with other scenarios, we have by Eq. (A-71), a measured deviation δ and a calibration process error ε_{cal}:

$$\begin{aligned} \delta &= e_{UUT,b} - e_{MTE,b} + (\varepsilon_{UUT,m} - \varepsilon_{MTE,m}) \\ &= e_{UUT,b} + \varepsilon_{cal}, \end{aligned}$$

(A-72)

where

$$\varepsilon_{cal} = (\varepsilon_{UUT,m} - \varepsilon_{MTE,m}) - e_{MTE,b}.$$

(A-73)

Letting $\epsilon_{c,b}$ represent the bias of the comparator, $\varepsilon_{MTE,m}$ is given by

$$\varepsilon_{MTE,m} = e_{c,b} + e_{MTE,rep} + e_{MTE,res} + e_{MTE,op} + e_{MTE,other}.$$

(A-74)

and $\varepsilon_{UUT,m}$ is

$$\varepsilon_{UUT,m} = e_{c,b} + e_{UUT,rep} + e_{UUT,res} + e_{UUT,op} + e_{UUT,other}.$$

(A-75)

By Eqs. (A-72) and (A-38), the measurement uncertainty in δ is obtained from

$$u_{cal} = \sqrt{\mathrm{var}(\varepsilon_{cal})},$$

where

$$\mathrm{var}(\varepsilon_{cal}) = u_{MTE,b}^2 + u_{rep}^2 + u_{res}^2 + u_{op}^2 + u_{other}^2.$$

(A-76)

In this scenario,

$$\begin{aligned} u_{MTE,b}^2 &= \mathrm{var}(-e_{MTE,b}) \\ u_{rep}^2 &= \mathrm{var}(e_{UUT,rep} - e_{MTE,rep}) = u_{MTE,rep}^2 + u_{UUT,rep}^2 \\ u_{res}^2 &= \mathrm{var}(e_{UUT,res} - e_{MTE,res}) = u_{MTE,res}^2 + u_{UUT,res}^2 \\ u_{op}^2 &= \mathrm{var}(e_{UUT,op} - e_{MTE,op}) = u_{MTE,op}^2 + u_{UUT,op}^2 - 2\rho_{op} u_{MTE,op} u_{UUT,op} \end{aligned}$$

(A-77)

and

$$u_{other}^2 = \mathrm{var}(e_{UUT,other} - e_{MTE,other}) = u_{MTE,other}^2 + u_{UUT,other}^2 - 2\rho_{other} u_{MTE,other} u_{UUT,other},$$

(A-78)

where ρ_{other} represents the correlation, if any, between $e_{MTE,other}$ and $e_{UUT,other}$.

---

[43] To accommodate cases where $y_n \neq x_n$, δ is redefined as

$$\delta = (x - x_n) - (y - y_n).$$

As an example where $x_n \neq y_n$, consider a case where the MTE is a 2 cm gage block and the UUT is a 1 cm gage block. Suppose that the comparator readings for the MTE and UUT are 2.10 cm and 0.99 cm, respectively. Then

$$\delta = (0.99 - 1.0) - (2.10 - 2.0) = -0.110 \text{ cm},$$

and, using Eq. (A-63), we have

$$x_c = 2.0 \text{ cm} + (0.99 - 2.10) \text{ cm} = (2.0 - 1.11) \text{ cm} = 0.89 \text{ cm}.$$

A.7.4 Scenario 4: MTE and UUT Measure a Common Attribute

In this scenario, both the MTE and UUT measure the value of an attribute of a common device or artifact, where the attribute provides an output or stimulus. The measurements are made and recorded separately. An example of this scenario is the calibration of a thermometer (UUT) using a temperature reference (MTE), where both the thermometer and the temperature reference are placed in an oven and the temperatures measured by each are recorded.

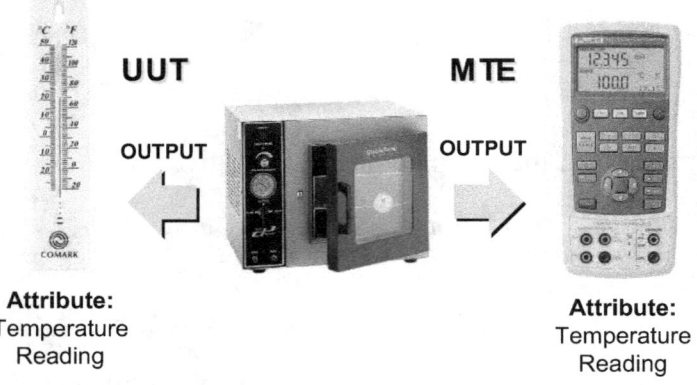

Figure A-11. Scenario 4

The UUT and the MTE Measure an Attribute of a Common Device or Artifact. The output is the temperature of an oven.

We let T denote the true value of the attribute and write the measurement equation as

$$x = T + \varepsilon_{UUT,m} , \qquad (A\text{-}79)$$

and

$$y = T + \varepsilon_{MTE,m} , \qquad (A\text{-}80)$$

where $\varepsilon_{UUT,m}$ is the measurement process error for the UUT temperature measurement and $\varepsilon_{MTE,m}$ is the measurement process error for the MTE temperature measurement. These errors are given by

$$\varepsilon_{UUT,m} = e_{UUT,b} + e_{UUT,rep} + e_{UUT,res} + e_{UUT,op} + e_{UUT,other} \qquad (A\text{-}81)$$

and

$$\varepsilon_{MTE,m} = e_{MTE,b} + e_{MTE,rep} + e_{MTE,res} + e_{MTE,op} + e_{MTE,other} . \qquad (A\text{-}82)$$

Substituting these expressions in Eqs. (A-79) and (A-80) gives

$$x = T + e_{UUT,b} + e_{UUT,rep} + e_{UUT,res} + e_{UUT,op} + e_{UUT,other} \qquad (A\text{-}83)$$

and

$$y = T + e_{MTE,b} + e_{MTE,rep} + e_{MTE,res} + e_{MTE,op} + e_{MTE,other} . \qquad (A\text{-}84)$$

Defining

$$\delta = x - y , \qquad (A\text{-}85)$$

these expressions yield

$$\delta = e_{UUT,b} + \varepsilon_{cal} . \qquad (A\text{-}86)$$

where

- 106 -

$$\varepsilon_{cal} = (e_{UUT,rep} - e_{MTE,rep}) + (e_{UUT,res} - e_{MTE,res})$$
$$+ (e_{UUT,op} - e_{MTE,op}) + (e_{UUT,other} - e_{MTE,other}) - e_{MTE,b}. \tag{A-87}$$

By Eq. (A-38), the measurement uncertainty is again given by

$$u_{cal} = \sqrt{\text{var}(\varepsilon_{cal})}, \tag{A-88}$$

where

$$\text{var}(\varepsilon_{cal}) = u_{MTE,b}^2 + u_{rep}^2 + u_{res}^2 + u_{op}^2 + u_{other}^2, \tag{A-89}$$

and

$$u_{MTE,b}^2 = \text{var}(-e_{MTE,b})$$
$$u_{rep}^2 = \text{var}(e_{UUT,rep} - e_{MTE,rep}) = u_{UUT,rep}^2 + u_{MTE,rep}^2$$
$$u_{res}^2 = \text{var}(e_{UUT,res} - e_{MTE,res}) = u_{UUT,res}^2 + u_{MTE,res}^2 \tag{A-90}$$
$$u_{op}^2 = \text{var}(e_{UUT,op} - e_{MTE,op}) = u_{UUT,op}^2 + u_{MTE,op}^2 - 2\rho_{op}u_{UUT,op}u_{MTE,op},$$

and

$$u_{other}^2 = \text{var}(e_{UUT,other} - e_{MTE,other}) = u_{UUT,other}^2 + u_{MTE,other}^2 - 2\rho_{other}u_{UUT,other}u_{MTE,other}, \tag{A-91}$$

where, again, ρ_{other} represents a correlation between $e_{MTE,other}$ and $e_{UUT,other}$.

A.7.4.1   Scenario 4 Special Cases

There are two special cases of Scenario 4 that may be thought of as variations of Scenarios 1 and 2. Both cases are accommodated by the Scenario 4 definitions and expressions developed above.

Case 1: The MTE measures the UUT and both the MTE and UUT provide a metered or other displayed output.

In this case, the common attribute is a "stimulus" embedded in the UUT. An example would be a UUT voltage source whose output is indicated by a digital display and is measured using an MTE voltmeter.

Case 2: The UUT measures the MTE and both the MTE and UUT provide a metered or other displayed output.

In this case, the common attribute is a "stimulus" embedded in the MTE. An example would be an MTE voltage source whose output is indicated by a digital display and is measured using a UUT voltmeter.

A.8   Uncertainty Analysis Summary

Four scenarios have been discussed that yield expressions for calibration uncertainty that are useful for risk analysis. In all scenarios and cases, the calibration result is expressed as

$$\delta = e_{UUT,b} + \varepsilon_{cal},$$

and the calibration uncertainty is given by

$$u_{cal} = \sqrt{\text{var}(\varepsilon_{cal})}.$$

Scenario 1: MTE Measures the UUT Attribute Value

In this scenario, the measurement result is $\delta = y - x_n$, and ε_{cal} is given in Eq. (A-42). The quantity $\text{var}(\varepsilon_{cal})$ is expressed in Eq. (A-47).

Scenario 2: UUT Measures the MTE Attribute Value

In this scenario, the measurement result is $\delta = x - y_n$ and ε_{cal} is given in Eq. (A-59). The quantity $\text{var}(\varepsilon_{cal})$ is expressed in Eq. (A-61).

Scenario 3: MTE and UUT Each Provide an Output

In this scenario, the measured UUT attribute value receives a correction given by

$$x_c = y_n + (x - y),$$

the measurement result is $\delta = x - y$, and ε_{cal} is given in Eq. (A-73). The quantity $\text{var}(\varepsilon_{cal})$ is expressed in Eq. (A-76).

Scenario 4: MTE and UUT Measure a Common Attribute

For this scenario, the measurement result is $\delta = x - y$ and ε_{cal} is given in Eq. (A-87). The quantity $\text{var}(\varepsilon_{cal})$ is expressed in Eq. (A-89).

Uncertainty Analysis Examples

Examples of uncertainty analyses for the four scenarios described in this appendix are given in [A-1] and [A-2]. In all scenarios and cases, the calibration result is expressed as

$$\delta = e_{UUT,b} + \varepsilon_{cal},$$

and the calibration uncertainty is given by

$$u_{cal} = \sqrt{\text{var}(\varepsilon_{cal})}.$$

Appendix A References

[A-1]   NCSLI, *Determining and Reporting Measurement Uncertainties*, Recommended Practice RP-12, NCSL International, Under Revision.

[A-2]   ANSI/NCSL, Z540.3-2006, *Requirements for the Calibration of Measuring and Test Equipment*, August 2006.

[A-3]   ANSI/ISO/IEC 17025-2000, *General Requirements for the Competence of Testing and Calibration Laboratories*, November 2000.

[A-4]   ANSI/NCSL Z540-2-1997, *U.S. Guide to the Expression of Uncertainty in Measurement*, October 1997.

[A-5]   Abramowitz, M, and Stegun, I, *Handbook of Mathematical Functions*, U.S. Dept. of Commerce Applied Mathematics Series **55**, 1972. Also available from www.dlmf.nist.gov.

Appendix B - Test and Calibration Quality Metrics

B.1   Introduction

False accept and false reject risk are metrics by which to measure the quality of a test or calibration process relative to the attributes tested or calibrated. As will be discussed in this appendix, there are several other metrics that can be employed. The choices that should be adopted as "standard" are those that are both relevant to the measurement community and consistent with ISO/IEC 17025 [B-1]. Consistence with ISO/IEC 17025 will be ensured if the metrics used are informative to the user of tested or calibrated attributes. Accordingly, selection of the risk metric should address, among other things, the organization's quality policy, any contractual requirements and any national, international or industry standards being used.

For discussion purposes, a tested or calibrated attribute of an item or system is assumed to have a specification consisting of a nominal or declared value and a tolerance limit, if the specification is single-sided, or an upper and lower tolerance limit if the specification is two-sided.

B.2   Appendix B Nomenclature

Let $e_{UUT,b}$ represent the deviation from nominal or bias of an attribute of interest and let δ represent a measurement of $e_{UUT,b}$. We define a performance or specification region L as all values of $e_{UUT,b}$ that lie within the attribute's tolerance limits. Similarly, we define an acceptance region $\mathcal{A}$ as all values of δ that lie within the attribute's acceptance limits.

We also define the following terms that will be useful in establishing measurement quality metrics:[44]

Table B-1. Measurement Quality Metrics Variables.

Variable		Description
$e_{UUT,b}$	-	variable representing the bias of the UUT attribute being tested or calibrated
$u_{UUT,b}$	-	standard uncertainty in $e_{UUT,b}$
δ	-	variable representing measurements of $e_{UUT,b}$
u_{cal}	-	standard uncertainty in δ
n	-	number of attributes tested or calibrated
n_g	-	number of in-tolerance ("good") attributes tested or calibrated
n_b	-	number of out-of-tolerance ("bad") attributes tested or calibrated
n_a	-	number of attributes accepted by testing or calibration
n_r	-	number of attributes rejected by testing or calibration
n_{ga}	-	number of in-tolerance attributes accepted by testing or calibration
n_{gr}	-	number of in-tolerance attributes rejected by testing or calibration
n_{ba}	-	number of out-of-tolerance attributes accepted by testing or calibration
n_{br}	-	number of out-of-tolerance attributes rejected by testing or calibration
UFAR	-	unconditional false accept risk
CFAR	-	conditional false accept risk
FRR	-	false reject risk
CFAR'	-	fraction of out-of-tolerance attributes that will be accepted
CFRR'	-	fraction of in-tolerance attributes that will be rejected
GA	-	fraction of in-tolerance attributes that will be accepted

---

[44] In this appendix, the variables x and y are substituted for the variables $e_{UUT,b}$ and δ of Appendix A for simplicity of notation.

Variable		Description
BR	-	fraction of out-of-tolerance attributes that will be rejected
CGA	-	probability of accepting in-tolerance attributes
CBR	-	probability of rejecting out-of-tolerance attributes
$-L_1$	-	lower tolerance limit for x
L_2	-	upper tolerance limit for x
L	-	the region $[-L_1, L_2]$
$P(x \in \text{L})$	-	probability that a tested or calibrated attribute will be in-tolerance
$-A_1$	-	lower acceptance limit for y
A_2	-	upper acceptance limit for y
$\mathcal{A}$	-	the region $[-A_1, A_2]$
$P(y \in \mathcal{A})$	-	probability that a tested or calibrated attribute will be accepted as being in-tolerance
$P(x \in \text{L}, y \in \mathcal{A})$	-	probability that a tested or calibrated attribute will both be in-tolerance and accepted by testing or calibration
$P(x \notin \text{L} \mid y \in \mathcal{A})$	-	probability that an attribute accepted by testing or calibration will be out-of-tolerance
$P(x \notin \text{L}, y \notin \mathcal{A})$	-	probability that an attribute will both be out-of-tolerance and rejected by testing or calibration.
$P(x \notin \text{L}, y \in \mathcal{A})$	-	probability that an attribute will both be out-of-tolerance and accepted by testing or calibration.

B.3 Discussion

B.3.1 Unconditional False Accept Risk

As pointed out in Chapters 3 and 4, the definition of false accept risk commonly encountered in measurement decision risk analysis articles and papers is the unconditional false accept risk *UFAR*, given by

$$UFAR = P(e_{UUT,b} \notin \text{L}, \delta \in \mathcal{A}).$$ (B-1)

To get some perspective on this definition of false accept risk, we first construct relationships that make use of the experimental definition of probability. That is, we claim that, as n becomes large,

$$n_a = P(\delta \in \mathcal{A})n,$$ (B-2)

and

$$n_{ba} = P(e_{UUT,b} \notin \text{L}, \delta \in \mathcal{A})n.$$ (B-3)

From Eqs. (B-1) and (B-3), we see that *UFAR* can be written

$$UFAR = \frac{n_{ba}}{n}.$$ (B-4)

This is the number of accepted out-of-tolerance attributes divided by the number of attributes tested or calibrated. This definition of false accept risk provides a metric that is relevant to the service provider. Its relevance to the equipment user is discussed in the next section.

B.3.2 Conditional False Accept Risk

The user is not typically interested in the number of out-of-tolerance attributes accepted relative to the lot of attributes tested or calibrated. From the user's perspective a more relevant variable is the number of attributes that are out-of-tolerance in the lot of attributes that were *accepted*.

Given this, if rejected attributes are not adjusted or otherwise corrected, and are not included in the accepted lot, a more relevant definition of false accept risk is[45]

$$CFAR = \frac{n_{ba}}{n_a}.$$

(B-5)

Using Eqs. (B-2) and (B-3), we have

$$CFAR = \frac{P(e_{UUT,b} \notin \mathsf{L}, \delta \in \mathcal{A})}{P(\delta \in \mathcal{A})},$$

(B-6)

which, by the rules of probability, can be written as the conditional probability

$$CFAR = P(e_{UUT,b} \notin \mathsf{L} \mid \delta \in \mathcal{A}).$$

(B-7)

Using the probability relations developed in Chapter 3, we have

$$P(\delta \in \mathcal{A}) = P(e_{UUT,b} \in \mathsf{L}, \delta \in \mathcal{A}) + P(e_{UUT,b} \notin \mathsf{L}, \delta \in \mathcal{A})$$

so that

$$P(e_{UUT,b} \notin \mathsf{L}, \delta \in \mathcal{A}) = P(\delta \in \mathcal{A}) - P(e_{UUT,b} \in \mathsf{L}, \delta \in \mathcal{A}).$$

Then Eq. (B-6) can be written

$$\begin{aligned} CFAR &= \frac{P(e_{UUT,b} \notin \mathsf{L}, \delta \in \mathcal{A})}{P(\delta \in \mathcal{A})} \\ &= \frac{P(\delta \in \mathcal{A}) - P(e_{UUT,b} \in \mathsf{L}, \delta \in \mathcal{A})}{P(\delta \in \mathcal{A})} \\ &= 1 - \frac{P(e_{UUT,b} \in \mathsf{L}, \delta \in A)}{P(\delta \in \mathcal{A})}. \end{aligned}$$

(B-8)

Note that we also can write

$$\begin{aligned} UFAR &= P(e_{UUT,b} \notin \mathsf{L}, \delta \in \mathcal{A}) \\ &= P(\delta \in \mathcal{A}) - P(e_{UUT,b} \in \mathsf{L}, \delta \in \mathcal{A}), \end{aligned}$$

---

[45] If rejected attributes are corrected in some way and subsequently returned to service, a more involved definition of *CFAR* is needed. This definition equates *CFAR* to the probability that UUT attributes will be out-of-tolerance following testing or calibration, regardless of whatever renewal action is taken. This probability is best computed using the "post-test distribution" discussed in Chapter 2 of the Handbook and in Annex 1.

It should be noted that, if the measurement uncertainty of the test or calibration process is much smaller than the tolerance limits of the UUT attribute, *UFAR* becomes a good approximation of the post-test *CFAR* computed in this way.

which shows that

$$CFAR = \frac{UFAR}{P(\delta \in \mathcal{A})}.$$

From this relation, we see that *CFAR* is a larger number than *UFAR*. Hence, computing false accept risk from perspective of the testing or calibration function may yield numbers that present a rosier picture than computations based on false accept risk from the user's perspective.

As an example, consider a situation in which 25% of the attributes in a lot of 1000 attributes are out-of-tolerance as received for testing. Imagine that the measuring system is such that there is a 6% probability that an attribute will be both out-of-tolerance and accepted as being in-tolerance:

$$P(e_{UUT,b} \notin \mathsf{L}, \delta \in \mathcal{A}) = 0.06.$$

In cases where the tested attributes have two-sided tolerances and have values that are normally distributed, these numbers correspond to an acceptance probability $P(\delta \in \mathcal{A})$ of about 68.7%, assuming that the acceptance region is synonymous with the performance region, i.e., $\mathcal{A} = \mathsf{L}$. Hence, we expect in this case that about 687 attributes will be accepted, and the equipment user will receive a number of tested attributes in which $60 / 687 \cong 8.7\%$ are out-of-tolerance.

If we employ the *UFAR* definition of false accept risk the reported risk will be only 6% — which an unwitting user might find acceptable. If, however, we the *CFAR* definition is applicable, the reported risk will be nearly 9%.

B.3.3   False Reject Risk

False reject risk is a quantity that is directly relevant to the test or calibration service provider and indirectly relevant to the user, to the extent that its value affects the cost of testing or calibration. From both the service provider's and user's viewpoint, false reject risk can be defined as the probability that an attribute will both be in-tolerance and rejected. Thus

$$\begin{aligned} FRR &= P(e_{UUT,b} \in \mathsf{L}, \delta \notin \mathcal{A}) \\ &= P(e_{UUT,b} \in \mathsf{L}) - P(e_{UUT,b} \in \mathsf{L}, \delta \in \mathcal{A}). \end{aligned} \tag{B-9}$$

Continuing with the above example, $P(e_{UUT,b} \in \mathsf{L}) = 0.75$ and

$$\begin{aligned} P(e_{UUT,b} \in \mathsf{L}, \delta \in A) &= P(\delta \in A) - P(e_{UUT,b} \notin \mathsf{L}, \delta \in A) \\ &= 0.687 - 0.06 \\ &= 0.627, \end{aligned}$$

so that $FRR = 0.75 - 0.627 = 0.123$. With a false reject risk of over 12%, we would expect some corrective measures would be sought. When we recall that this example is an instance where the commonly defined false accept risk is only 6%, these measures might not be forthcoming if the service provider is not cognizant of false reject risk.

B.3.4   Other Metrics

From the foregoing, it is clear that three metrics of relevance to both user and service provider are *UFAR*, *CFAR* and *FRR*, as defined in Eqs. (B-1), (B-7) and (B-9). In addition to these, there are others that could be of interest to quality managers and metrologists. For example, we might be interested in the fraction of out-of-tolerance attributes that will be accepted and the fraction of in-tolerance attributes that will be rejected. These metrics are respectively given by

$$
\begin{aligned}
CFAR' &= \frac{n_{ba}}{n_b} \\
&= \frac{P(e_{UUT,b} \notin \mathsf{L}, \delta \in \mathcal{A})}{P(e_{UUT,b} \notin \mathsf{L})} = P(\delta \in \mathcal{A} \mid e_{UUT,b} \notin \mathsf{L}) \\
&= \frac{P(\delta \in \mathcal{A}) - P(e_{UUT,b} \in \mathsf{L}, \delta \in \mathcal{A})}{1 - P(e_{UUT,b} \in \mathsf{L})},
\end{aligned}
\tag{B-10}
$$

and

$$
\begin{aligned}
FRR' &= \frac{n_{gr}}{n_g} \\
&= \frac{P(e_{UUT,b} \in \mathsf{L}, \delta \notin \mathcal{A})}{P(e_{UUT,b} \in \mathsf{L})} = P(\delta \notin \mathcal{A} \mid e_{UUT,b} \in \mathsf{L}) \\
&= \frac{P(e_{UUT,b} \in \mathsf{L}) - P(e_{UUT,b} \in \mathsf{L}, \delta \in \mathcal{A})}{P(e_{UUT,b} \in \mathsf{L})} \\
&= 1 - \frac{P(e_{UUT,b} \in \mathsf{L}, \delta \in \mathcal{A})}{P(e_{UUT,b} \in \mathsf{L})}.
\end{aligned}
\tag{B-11}
$$

Continuing with the foregoing example, we have

$$FAR' = 0.06 / 0.25 = 0.24,$$

and

$$FRR' = 1 - 0.627 / 0.75 = 0.164.$$

These metrics show that 24% of out-of-tolerance attributes will be accepted and over 16% of in-tolerance attributes will be rejected by the testing or calibration process. While such numbers would be of interest primarily to the testing or calibration organization, it is easy to see that they could be useful in identifying measurement quality problems.

Additional metrics are possible. For instance, we might be interested in the probability of accepting in-tolerance attributes and the probability of rejecting out-of-tolerance attributes.

The probability of accepting in-tolerance attributes is given by

$$
\begin{aligned}
CGA &= \frac{n_{ga}}{n_g} \\
&= \frac{P(e_{UUT,b} \in \mathsf{L}, \delta \in \mathcal{A})}{P(e_{UUT,b} \in \mathsf{L})} = P(\delta \in \mathcal{A} \mid e_{UUT,b} \in \mathsf{L}).
\end{aligned}
\tag{B-12}
$$

With our example, we have

$$CGA = 0.627 / 0.75 \cong 0.836.$$

The probability of rejecting out-of-tolerance attributes is expressed as

$$CBR = \frac{n_{br}}{n_b}$$

$$= \frac{P(e_{UUT,b} \notin \mathsf{L}, \delta \notin \mathcal{A})}{1 - P(e_{UUT,b} \in \mathsf{L})} = P(\delta \notin \mathcal{A} \mid e_{UUT,b} \notin \mathsf{L}).$$

(B-13)

To compute the numerator using numbers at our disposal, we first need to decompose $P(e_{UUT,b} \notin \mathsf{L}, \delta \notin \mathcal{A})$. Using the probability relations developed in Chapter 3, we can write

$$P(e_{UUT,b} \notin \mathsf{L}, \delta \notin \mathcal{A}) = 1 - P(e_{UUT,b} \in \mathsf{L} \text{ or } \delta \in \mathcal{A}).$$

But $P(e_{UUT,b} \in \mathsf{L} \text{ or } \delta \in \mathcal{A})$ can be expressed as

$$P(e_{UUT,b} \in \mathsf{L} \text{ or } \delta \in \mathcal{A}) = P(e_{UUT,b} \in \mathsf{L}) + P(\delta \in \mathcal{A}) - P(e_{UUT,b} \in \mathsf{L}, \delta \in \mathcal{A}).$$

Then

$$P(e_{UUT,b} \notin \mathsf{L}, \delta \notin \mathcal{A}) = 1 + P(e_{UUT,b} \in \mathsf{L}, \delta \in \mathcal{A}) - P(e_{UUT,b} \in \mathsf{L}) - P(\delta \in \mathcal{A}),$$

which yields

$$CBR = 1 - \frac{P(\delta \in \mathcal{A}) - P(e_{UUT,b} \in \mathsf{L}, \delta \in \mathcal{A})}{1 - P(e_{UUT,b} \in \mathsf{L})}$$

$$= 1 - \frac{P(e_{UUT,b} \notin \mathsf{L}, \delta \in \mathcal{A})}{1 - P(\delta \in \mathsf{L})}.$$

(B-14)

With our example, we have

$$CBR = 1 - 0.06 / 0.25 \cong 0.76.$$

These metrics indicate that about 84% of in-tolerance attributes will be accepted, while 76% of out-of-tolerance attributes will be rejected. Again, these figures are relevant primarily to the service provider, but may prove useful as quality control variables.

B.4   Controlling Risks with Guardbands

False accept and false reject risks can be controlled at the process-level by the imposition of guardband limits [2], [3]. The development of such limits involves additional probability relations.

B.4.1   Guardband Risk Relations

It is often useful to relate the range of acceptable values $\mathcal{A}$ to the range L by variables called **guardband multipliers**. Let g_1 and g_2 be lower and upper guardband multipliers, respectively. If $-L_1$ and L_2 are the lower and upper attribute tolerance limits, and $-A_1$ and A_2 the corresponding acceptance limits, then

$$A_1 = g_1 L_1$$
$$A_2 = g_2 L_2.$$

Suppose that g_1 and g_2 are both < 1. Then the acceptance region $\mathcal{A}$ is smaller than the performance region L. If $\mathcal{A}$ is a subset of L then

$$P(e_{UUT,b} \notin \mathsf{L}, \delta \in \mathcal{A}) < P(e_{UUT,b} \notin \mathsf{L}, \delta \in \mathsf{L}).$$

From the definition of *UFAR* in Eq. (B-1), we see that, if $\mathcal{A} < \mathsf{L}$, then the estimated value of *UFAR* is less than if $\mathcal{A} = \mathsf{L}$. Likewised, from the definition of *CFAR* in Eq. (B-7), it can be shown that, if $\mathcal{A} < \mathsf{L}$ then, then the estimated value of *CFAR* is also less than if $\mathcal{A} = \mathsf{L}$.

So, setting the guardband limits inside tolerance limits reduces false accept risk and increases false reject risk. Conversely, setting guardband limits outside tolerance limits increases false accept risk and reduces false reject risk.

B.4.2 Establishing Risk-Based Guardbands

Guardbands can often be set to achieve a desired level of false accept or false reject risk, *R*. Cases where it is not possible to establish guardbands are those where the desired level of risk is not attainable even with guardband limits set to zero. Moreover, solutions for guardbands are usually restricted to finding symmetric guardband multipliers, i.e., those for which $g_1 = g_2 = g$.

B.4.2.1 False Accept Risk-Based Guardbands

False accept risk-based guardbands are established numerically by iteration.[46] The iteration adjusts the value of a symmetric guardband multiplier *g*, until the false accept risk (*FAR*) is approximately equal to a maximum allowable risk *R*. The following algorithm illustrates the process:

Step 1: Set $g = 1$.
Step 2: Set $A_1 = gL_1$ and $A_2 = gL_2$
Step 3: Compute $P(\delta \in \mathcal{A})$ and $P(e_{UUT,b} \in \mathsf{L}, \delta \in \mathcal{A})$
Step 4: Compute false accept risk *FAR* (either *UFAR* or *CFAR*, as appropriate).
Step 5: If $FAR < R$, go to Case 1. If $FAR > R$, go to Case 2 | Double. If $FAR \cong R$, the process is complete.

Case 1: ***FAR < R***

> Set $g = g / 2$
> Repeat Steps 2 through 4
> If $FAR < R$, go to Case 1. If $FAR > R$, go to Case 2 | Bisect. If $FAR \cong R$, the process is complete.

Case 2: ***FAR > R***

> Double: Set $g = 2g$
> Repeat Steps 2 through 4
> If $FAR < R$, go to Case 1. If $FAR > R$, go to Case 2 | Double. If $FAR \cong R$, the process is complete.

---

[46] The bisection method of Appendix F is recommended.

Bisect: Set $g = g + g / 2$
Repeat Steps 2 through 4
If $FAR < R$, go to Case 1. If $FAR > R$, go to Case 2 | Bisect. If $FAR \cong R$, the process is complete.

B.4.2.2   False Reject Risk-Based Guardbands

False reject risk-based guardbands are established in the same way as false accept risk-based guardbands. The following algorithm illustrates the process:

Step 1: Set $g = 1$

Step 2: Set $A_1 = gL_1$ and $A_2 = gL_2$

Step 3: Compute $P(e_{UUT,b} \in L)$ and $P(e_{UUT,b} \in L, \delta \in \mathcal{A})$

Step 4: Compute false reject risk FRR

Step 5: If $FRR < R$, go to Case 1 | Double. If $FRR > R$, go to Case 2. If $FRR \cong R$, the process is complete.

Case 1: $FR < R$

Double: Set $g = 2g$
Repeat Steps 2 through 4
If $FRR < R$, go to Case 1 | Double. If $FRR > R$, go to Case 2. If $FRR \cong R$, the process is complete.

Bisect: Set $g = g + g / 2$
Repeat Steps 2 through 4
If $FRR < R$, go to Case 1 | Bisect. If $FRR > R$, go to Case 2. If $FRR \cong R$, the process is complete.

Case 2: $FR > R$

Set $g = g / 2$
Repeat Steps 2 through 4
If $FRR < R$, go to Case 1 | Bisect. If $FRR > R$, go to Case 2. If $FRR \cong R$, the process is complete.

B.5   Computing Probabilities

In computing measurement decision risks, we use uncertainty estimates, along with *a priori* information, to compute the probabilities needed to estimate false accept and false reject risks.[47]

B.5.1   The Basic Set of Integrals

The in-tolerance probability $P(e_{UUT,b} \in L)$ is written

$$P(e_{UUT,b} \in L) = \int_L f(e_{UUT,b}) de_{UUT,b} \cdot$$

Let the joint pdf of $e_{UUT,b}$ and δ be denoted $f(e_{UUT,b}, \delta)$. Then the probability that the UUT attribute is both in-tolerance and observed to be in-tolerance is given by

---

[47] The subject of computing probabilities is also covered in Section 4.2.2 using the notation of the calibration scenarios described in Appendix A.

$$P(e_{UUT,b} \in \mathsf{L}, \delta \in \mathcal{A}) = \int_{\mathsf{L}} de_{UUT,b} \int_{\mathcal{A}} f(e_{UUT,b}, \delta) d\delta$$

$$= \int_{\mathsf{L}} f(e_{UUT,b}) de_{UUT,b} \int_{\mathcal{A}} \hat{f}(\delta \mid e_{UUT,b}) d\delta,$$

where the function $f(\delta|e_{UUT,b})$ is the conditional pdf of obtaining a value (measurement) δ, given that the bias of the value being measured is $e_{UUT,b}$.

B.5.1.1 Equipment Attribute Distributions

Reference Attribute

The normal distribution is usually assumed for the variable δ, conditional on the value of the variable $e_{UUT,b}$. Accordingly, the reference attribute pdf is given by

$$f(\delta \mid e_{UUT,b}) = \frac{1}{\sqrt{2\pi}u_{cal}} e^{-(\delta - e_{UUT,a})^2/2u_{cal}^2},$$

where u_{cal} is the standard uncertainty in the measurement, given by

$$u_{cal} = \sqrt{u_{MTE,b}^2 + u_{other}^2}.$$

In this expression, $u_{MTE,b}$ is the uncertainty in the reference attribute bias and u_{other} is the combined standard uncertainty for any remaining measurement process errors. The determination of u_{cal} is covered in detail in Appendix A

UUT Attribute

A few useful UUT attribute distributions are described in Section A.2.2.2, in Appendix E and in the literature [B-4]. For purposes of illustration, the following employs the normal distribution. With this distribution, the pdf for $e_{UUT,b}$ is given by

$$f(e_{UUT,b}) = \frac{1}{\sqrt{2\pi}u_{UUT,b}} e^{-e_{UUT,b}^2/2u_{UUT,b}^2},$$

where $u_{UUT,b}$ is the *a priori* standard deviation of the UUT attribute bias population.[48] For this pdf, the function $P(e_{UUT,b} \in \mathsf{L})$ becomes

$$P(e_{UUT,b} \in \mathsf{L}) = \frac{1}{\sqrt{2\pi}u_x} \int_{\mathsf{L}} e^{-e_{UUT,b}^2/2u_y^2} de_{UUT,b}.$$

Since measurements δ of $e_{UUT,b}$ follow a normal distribution with standard deviation u_y and mean equal to $e_{UUT,b}$, the function $P(e_{UUT,b} \in \mathsf{L}, \delta \in \mathcal{A})$ is written as

$$P(e_{UUT,b} \in \mathsf{L}, \delta \in \mathcal{A}) = \frac{1}{2\pi u_{UUT,b} u_{cal}} \int_{\mathsf{L}} e^{-e_{UUT,b}^2/2u_{UUT,b}^2} de_{UUT,b} \int_{\mathcal{A}} e^{-(\delta - e_{UUT,b})^2/2u_{cal}^2} d\delta.$$

The function $P(y \in \mathcal{A})$ is obtained by integrating $f(e_{UUT,b}, y)$ over all values of $e_{UUT,b}$

---

[48] See Appendix B of Annex 1 or Appendix B of Annex 3 for expressions used to obtain estimates of the *a priori* or "pre-test" value of $u_{UUT,b}$.

$$P(\delta \in \mathcal{A}) = \frac{1}{2\pi u_{UUT,b} u_{cal}} \int_{-\infty}^{\infty} e^{-e_{UUT,b}^2 / 2u_{UUT,b}^2} de_{UUT,b} \int_{\mathcal{A}} e^{-(\delta - e_{UUT,b})^2 / 2u_{cal}^2} d\delta$$

$$= \frac{1}{\sqrt{2\pi} u_A} \int_{\mathcal{A}} e^{-\delta^2 / 2u_A^2} d\delta,$$

where

$$u_A = \sqrt{u_{UUT,b}^2 + u_{cal}^2}.$$

Consider the case of symmetric tolerance limits and guardband limits. Then the tolerance region can be expressed as $\pm L$ and the acceptance region as $\pm A$, then the above expressions become

$$P(e_{UUT,b} \in \mathsf{L}) = \frac{1}{\sqrt{2\pi} u_{UUT,b}} \int_{-L}^{L} e^{-e_{UUT,b}^2 / 2u_{UUT,b}^2} de_{UUT,b}$$

$$= 2\Phi\left(L / u_{UUT,b}\right) - 1,$$

$$P(e_{UUT,b} \in \mathsf{L}, \delta \in \mathcal{A}) = \frac{1}{2\pi u_{UUT,b} u_{cal}} \int_{-L}^{L} e^{-e_{UUT,b}^2 / 2u_{UUT,b}^2} de_{UUT,b} \int_{-A}^{A} e^{-(\delta - e_{UUT,b})^2 / 2u_{cal}^2} d\delta$$

$$= \frac{1}{\sqrt{2\pi} u_{UUT,b}} \int_{-L}^{L} \left[\Phi\left(\frac{A - e_{UUT,b}}{u_{cal}}\right) - \Phi\left(-\frac{A + e_{UUT,b}}{u_{cal}}\right) \right] e^{-e_{UUT,b}^2 / 2u_{UUT,b}^2} de_{UUT,b}$$

$$= \frac{1}{\sqrt{2\pi}} \int_{-L/u_{UUT,b}}^{L/u_{UUT,b}} \left[\Phi\left(\frac{A - u_{UUT,b}\zeta}{u_{cal}}\right) + \Phi\left(\frac{A + u_{UUT,b}\zeta}{u_{cal}}\right) - 1 \right] e^{-\zeta^2 / 2} d\zeta,$$

and

$$P(\delta \in \mathcal{A}) = \frac{1}{\sqrt{2\pi} u_A} \int_{-A}^{A} e^{-y^2 / 2u_A^2} d\delta$$

$$= 2\Phi\left(A / u_A\right) - 1.$$

The value of the joint probability $P(e_{UUT,b} \in \mathsf{L}, \delta \in \mathcal{A})$ is obtained by numerical integration.

B.6   Measurement Quality Metrics Estimation Procedure

The procedure for calculating the measurement quality metrics defined in this appendix follows the basic "recipe" shown below.

1. Establish the relevant quantities
    ▸ The *a priori* UUT attribute distribution.
    ▸ The UUT attribute in-tolerance probability prior to test or calibration.
    ▸ An estimate of the UUT attribute bias at the time of test or calibration.
    ▸ The reference attribute distribution.
    ▸ The reference attribute tolerances.
    ▸ The in-tolerance probability for the reference attribute at the time of test or calibration.
    ▸ An estimate of the reference attribute bias at the time of test or calibration.

2. Estimate the measurement process uncertainty.

3. Compute the metric *UFAR*, *CFAR*, *FRR*, *CFAR'*, *CFRR'*, *GA* or *BR*.

4. Evaluate the metric to determine if corrective action is needed.

B.7   Conclusion

Which metric or metrics to use to evaluate measurement quality is a matter of choice within the context of the objectives of the test or calibration function. The most common are *UFAR* or *CFAR* and *FRR*. *CFAR* is applicable in cases where rejected attributes are not corrected and subsequently accepted. *UFAR* is applicable in cases where the UUT in-tolerance probability is high, the TUR is large and rejected attributes are corrected and subsequently accepted.[49]

For cases in between, either the Bayesian analysis method or the methods described in Handbook Chapter 2 and in Annex 2 are recommended.

Use of the metrics *FRR*, *CFAR'* and *CFRR'* *GA* and *BR* should be governed by relevance to the service provider or equipment user.

Note:

Each metric described in this appendix has its value depending on the test and calibration scenario and the needs of the observer. Each is representative of an unperturbed test and calibration process where no actions are taken with rejected items. When adjustments or rework to the attributes of these items occurs, additional computation is required to evaluate the resulting effects as adjusted attributes have their own error distributions and risk parameters. If these are "mixed" with accepted attributes, for example, the risk factors change and should be taken into consideration for evaluation of the relative merit of the risk parameters.

Appendix B References

[B-1] ISO/IEC 17025 1999(e), *General Requirements for the Competence of Testing and Calibration Laboratories*, ISO/IEC, December 1999.

[B-2] Hutchinson, B., "Setting Guardband Test Limits to Satisfy MIL-STD-45662A Requirements," *Proc. NCSL Workshop & Symposium*, August 1991, Albuquerque.

[B-3] Deaver, D., "Guardbanding with Confidence," *Proc. 1994 NCSL Workshop and Symposium*, Chicago, July - August 1994.

[B-4] Castrup, H., "Selecting and Applying Error Distributions in Uncertainty Analysis," *Proc. Meas. Sci. Conf.*, Anaheim, January 2004. Available from www.isgmax.com.

[B-5] ANSI/NCSL Z540.3-2006, *Requirements for the Calibration of Measuring and Test Equipment*, August 2006.

---

[49] No nominal guidelines can be given as to what constitutes "high" and "large." These measures depend on the specifics of the test or measurement and on the risk control objectives.

Note that Z540.3 [B-5] requires that *UFAR* be two percent or less.

Appendix C: Introduction to Bayesian Measurement Decision Risk Analysis

In the 18th century, Reverend Thomas Bayes[50] expressed the probability of any event – given that a related event has occurred – as a function of the probabilities of the two events occurring independently and the probability of both events occurring together. The expression derived by Bayes is referred to as "Bayes' theorem."[51]

As we will see, Bayes' theorem has profound implications for the way we make decisions based on tests or measurements. To illustrate, take the example in which a patient sees a doctor for a checkup. The doctor knows a test he performs to diagnose a specific illness is 99 percent reliable – that is, 99 percent of sick people test positive, and 99 percent of healthy people test negative. The doctor also knows that only 1 percent of the general population is sick.[52]

Imagine that the patient tests positive. The doctor knows the chance of testing positive if the patient is sick, but what the patient wants to know is the chance that he is sick if he tests positive. The intuitive answer is 99 percent, but the correct answer is 50 percent. To arrive at this answer, we need to use Bayes' theorem.

Bayes' theorem emerges from the mathematics of probability. In keeping with the customary notation of this discipline, we denote probability with the letter p and the probability of some event E taking place as $p(E)$. In the present example, we will denote the event that the patient is sick by the letter s and the probability of obtaining a positive test result by the + symbol. Hence, the probability of being sick is written $p(s)$ and the probability of obtaining a positive result is written $p(+)$.

There are two more probability relations that need to be described; one is the joint probability of two events occurring together and the other is the conditional probability that one event will occur, given that another event has occurred. With joint probabilities, the two events are separated by a comma. With this convention, the probability of being sick *and* getting a positive test result is written $p(s,+)$. For conditional probability, the first event and the subsequent event are separated by the | symbol. With this convention, the probability of getting a positive test result, *given* that the patient is sick is written $p(+|s)$.

What we are interested in solving for is the probability of being sick, given a positive test result $p(s|+)$. We solve for probability through the use of Bayes' theorem.

To get to Bayes' theorem, we start by stating a powerful and simple relationship between joint probabilities and conditional probabilities. For example, the relationship between $p(+,s)$ and $p(+|s)$ is

---

[50] Bayes was born in 1702, was Presbyterian Minister at Tunbridge Wells from before 1731 until 1752, and died in 1761. For more information, see *Biometrica* **45**, 1958, 293-315.

[51] *Phil. Trans.* **53**, 1763, 376-98. Evidently, the paper was published posthumously.

[52] This example is adapted from C. Wiggins, "How can Bayes' theorem assign a probability to the existence of God?," *Scientific American*, April 2007.

$$p(+,s) = p(+|s)p(s). \tag{C-1}$$

Likewise, we have

$$p(s,+) = p(s|+)p(+). \tag{C-2}$$

Of course, the joint probability of getting both a positive result and being sick is the same as the joint probability of being sick and getting a positive result, i.e., $p(s,+) = p(+,s)$. So, by Eqs. (C-1) and (C-2), we can write

$$p(s|+)p(+) = p(+|s)p(s). \tag{C-3}$$

Dividing both sides by $p(+)$ yields Bayes' theorem

$$p(s|+) = \frac{p(+|s)p(s)}{p(+)}. \tag{C-4}$$

From the information we have been given in the example, we note that $p(s) = 0.01$ and $p(+|s) = 0.99$. If we indicate the event in which the patient is not sick by $\bar{s}$, we also know that $p(+|\bar{s}) = 0.01$. What we need now is the probability $p(+)$.

We first observe that there are only two ways to obtain a positive test result – either we obtain a positive result with a sick patient or with a healthy patient. In the language of probability theory, this is written as

$$p(+) = p(+,s) + p(+,\bar{s}). \tag{C-5}$$

We already have $p(+,s)$ in Eq. (C-1). Similarly, we can write the joint probability $p(+,\bar{s})$ of having a healthy patient and a positive test result as

$$\begin{aligned} p(+,\bar{s}) &= p(+|\bar{s})p(\bar{s}) \\ &= p(+|\bar{s})[1 - p(s)]. \end{aligned} \tag{C-6}$$

Combining Eqs. (C-1) and (C-5) in Eq. (C-5), yields

$$p(+) = p(+|s)p(s) + p(+|\bar{s})[1 - p(s)], \tag{C-7}$$

and plugging in the numbers gives

$$\begin{aligned} p(+) &= (0.99)(0.01) + (0.01)(0.99) \\ &= 2(0.99)(0.01). \end{aligned}$$

This result, together with the values we have for $p(+|s)$ and $p(s)$, when substituted in Eq. (C-4), gives

$$p(s|+) = \frac{(0.99)(0.01)}{2(0.99)(0.01)} = 0.5.$$

It is apparent that Bayes' theorem can have practical implications – in this case, a decision whether to recommend a specific treatment or to pursue some alternative course.

It is interesting to examine the impact that $p(s)$ can have on decisions made on the outcome of the test. For example, suppose all the numbers are the same as above, except that there is a 0.5 percent chance of being sick. Then $p(s|+)$ turns out to be 0.332 or a little over 33 percent. As another example, suppose that the probability of being sick is 2 percent. Then $p(s \mid +)$ turns out to be 0.669 or almost 67 percent. Evidently, the effect of $p(s)$ on $p(s \mid +)$ is quite dramatic. This is shown in Table C-1.

Table C-1. Bayes' Theorem Results for Different Values of $p(s)$.[53]

% of People Who are Sick $p(s)$	% Chance of Being Sick if Tested Positive $p(s \mid +)$
0.2	16.6
0.5	33.2
1.0	50.0
2.0	66.9
5.0	83.9

C.1   Application to Measurement Decision Risk Analysis

C.1.1   Conditional False Accept Risk

In estimating the probability of making false decisions in calibration and testing, we want to calculate the probability that a unit-under-test (UUT) attribute, accepted as in-tolerance by the calibration or test process, is truly in-tolerance. To do this, we use Bayes' theorem to relate (1) the probability of an accepted UUT attribute being in-tolerance $p(in|accept)$ to (2) the probability of observing an in-tolerance, given that the UUT attribute is in-tolerance, $p(accept|in)$, (3) the general probability of the attribute being in-tolerance, $p(in)$, and (4) the probability of observing an in-tolerance $p(accept)$. With this notation, Bayes' theorem for the probability that accepted UUT attributes will be in-tolerance is written as

$$p(in \mid accept) = \frac{p(accept \mid in)p(in)}{p(accept)} . \tag{C-8}$$

From the relationship between joint and conditional probabilities discussed earlier, we can express the joint probability $p(accept, in)$ as

$$p(accept, in) = p(accept \mid in)p(in) . \tag{C-9}$$

We can also write the joint probability $p(accept, out)$ as

$$\begin{aligned} p(accept, out) &= p(accept \mid out)p(out) \\ &= p(accept \mid out)[1 - p(in)]. \end{aligned} \tag{C-10}$$

We know that the UUT attribute is either in- or out-of-tolerance. So, there are only two circumstances under which we accept the attribute. As in the medical example, we use the language of probability theory and write

$$p(accept) = p(accept, in) + p(accept, out) . \tag{C-11}$$

Combining Eqs. (C-9) and (C-10) in Eq. (C-11) yields

$$p(accept) = p(accept \mid in)p(in) + p(accept \mid out)[1 - p(in)]. \tag{C-12}$$

---

[53] Values are computed using Eq. (4) for examples in which $p(+ \mid s) = 0.99$ and $p(+ \mid \bar{s}) = 0.01$.

For the sake of discussion, assume that 85 percent of the time the UUT attribute of interest is received for calibration in-tolerance. Also, suppose there is a 2 percent chance of accepting an out-of-tolerance attribute and a 98 percent chance of accepting an in-tolerance one. Then $p(accept \mid in) = 0.98$, $p(in) = 0.85$, $p(accept \mid out) = 0.02$, and Eq. (C-12) gives

$$p(accept) = (0.98)(0.85) + (0.02)[1 - (0.85)]$$
$$= 0.836.$$

Substituting the values for the constituent probabilities in Eq. (C-8) yields

$$p(in \mid accept) = \frac{(0.98)(0.85)}{0.836}$$
$$= 0.9964.$$

Thus, if no adjustments or other corrections are made to attributes that are observed to be in-tolerance, 99.64 percent of accepted attributes will be in-tolerance and only 0.36 percent will be out-of-tolerance. In other words, we will make an incorrect decision 0.36 percent of the time.

Note that we could have written Bayes' theorem as

$$p(out \mid accept) = \frac{p(accept \mid out)p(out)}{p(accept)}, \qquad (C-13)$$

and would have gotten

$$p(out \mid accept) = \frac{(0.02)(0.15)}{0.836} = 0.0036.$$

The function $p(out \mid accept)$ is called conditional false accept risk (*CFAR*).

C.1.2 Unconditional False Accept Risk

Another function that has been found to be useful is unconditional false accept risk (*UFAR*), defined as the joint probability of the attribute being out-of-tolerance and accepted as in-tolerance. *UFAR* is given by

$$p(out, accept) = p(accept \mid out)p(out)$$

In the present example,

$$p(out, accept) = (0.02)(0.15)$$
$$= 0.0030.$$

Notice that $UFAR < CFAR$. This is true in general.

C.1.3 False Reject Risk

From the foregoing, it is easy to develop a function giving the probability $p(in, reject)$ that an attribute will be both in-tolerance and rejected as being out-of-tolerance. This function is called false reject risk (*FRR*). To arrive at an expression for *FRR*, we first note that there are two outcomes for an in-tolerance attribute. Either it is accepted or rejected. Accordingly,

$$p(in) = p(in, accept) + p(in, reject), \qquad (C-14)$$

so that

$$p(in, reject) = p(in) - p(in, accept).$$

Since $p(accept, in) = p(in, accept)$, we can also write $p(in, reject)$ as

$$p(in, reject) = p(in) - p(accept, in). \tag{C-15}$$

Expressing the joint probability $p(accept, in)$ in terms of the conditional probability $p(accept|in)$, Eq. (C-15) can be expressed as

$$\begin{aligned} p(in, reject) &= p(in) - p(accept \mid in)p(in) \\ &= p(in)[1 - p(accept \mid in)]. \end{aligned} \tag{C-16}$$

For the example, $p(in) = 0.85$, $p(accept \mid in) = 0.98$, and Eq. (C-16) gives

$$\begin{aligned} p(in, reject) &= (0.85)(0.02) \\ &= 0.0170. \end{aligned}$$

C.1.4 Conditional False Reject Risk

Another useful function is conditional false reject risks (*CFRR*). This is the probability $p(reject \mid in)$ of rejecting an in-tolerance attribute. From the relationship between joint and conditional probabilities, we have

$$p(in, reject) = p(reject \mid in)p(in),$$

so that

$$p(reject \mid in) = \frac{p(in, reject)}{p(in)}. \tag{C-17}$$

This probability can also be written

$$p(reject \mid in) = 1 - p(accept \mid in). \tag{C-18}$$

For the present example, since $p(accept \mid in) = 0.98$,

$$p(reject \mid in) = 1 - 0.98 = 0.02.$$

This is the same answer as is obtained using Eq. (C-17).

C.2 Application of Bayes' Theorem with Measured Values

The utility of Bayes' theorem in estimating probabilities for false accepts and false rejects is not restricted to cases that make use of the probabilities $p(accept|in)$ and $p(accept|out)$. The real power of Bayes' theorem allows estimating false accept probabilities in cases where the result of calibration is a specific measured value – not just an in-tolerance or out-of-tolerance observation. For such cases, we avail ourselves of the foregoing probability definitions, but, instead of starting with probabilities, we work with probability density functions (pdfs). A pdf $f(x)$ is related to a probability $p(X)$ according to

$$p(X) = \int_{-\infty}^{X} f(x)dx. \tag{C-19}$$

This definition can be used to show that pdfs follow the same rules as probabilities. So, for two variables x and y, we have

$$f(x,y) = f(x|y)f(y) = f(y|x)f(x), \qquad \text{(C-20)}$$

and we can write Bayes' theorem as

$$f(x|y) = \frac{f(y|x)f(x)}{f(y)}. \qquad \text{(C-21)}$$

In general, calibration of a UUT attribute against a measurement reference attribute yields a calibration result δ, which is an estimate of the bias of the UUT attribute, denoted $e_{UUT,b}$. Various calibration scenarios have been identified [C-1], with expressions for δ and $e_{UUT,b}$ given for each.[54] Other relevant quantities are the uncertainty in the calibration process, denoted u_{cal}, and the bias uncertainty of the UUT as received for calibration, denoted $u_{UUT,b}$.

Using Eq. (C-21), Bayes' theorem for each scenario is given by

$$f(e_{UUT,b}|\delta) = \frac{f(\delta|e_{UUT,b})f(e_{UUT,b})}{f(\delta)}. \qquad \text{(C-22)}$$

Ordinarily, the pdf $f(\delta|e_{UUT,b})$ is considered to be normal and is written

$$f(\delta|e_{UUT,b}) = \frac{1}{\sqrt{2\pi}u_{cal}}e^{-(\delta-e_{UUT,b})/2u_{cal}^2}. \qquad \text{(C-23)}$$

The pdf $f(e_{UUT,b})$ may or may not be normal. Cases where a lognormal, uniform or other distribution is applicable have been noted [C-2] but are not common, and the normal distribution is usually assumed. Under this assumption,

$$f(e_{UUT,b}) = \frac{1}{\sqrt{2\pi}u_{UUT,b}}e^{-e_{UUT,b}^2/2u_{UUT,b}^2}. \qquad \text{(C-24)}$$

To obtain a Bayes' relation, we first need to develop the pdf $f(\delta)$. It can be shown that this is obtained by integrating the joint pdf $f(\delta, e_{UUT,b})$ over all possible values of $e_{UUT,b}$. By Eq. (C-20), we have

$$f(\delta, e_{UUT,b}) = f(\delta|e_{UUT,b})f(e_{UUT,b}),$$

and

---

[54] See Section 4.3 and Appendix A.

$$f(\delta) = \int\limits_{-\infty}^{\infty} f(\delta, e_{UUT,b}) de_{UUT,b}$$

$$= \int\limits_{-\infty}^{\infty} f(\delta \mid e_{UUT,b}) f(e_{UUT,b}) de_{UUT,b}$$

$$= \frac{1}{2\pi u_{cal} u_{UUT,b}} \int\limits_{-\infty}^{\infty} e^{-(\delta - e_{UUT,b})/2u_{cal}^2} e^{-e_{UUT,b}^2/2u_{UUT,b}^2} de_{UUT,b} \,.$$

After some rearranging, followed by completion of the integration, this reduces to

$$f(\delta) = \frac{1}{\sqrt{2\pi}u_A} e^{-\delta^2/2u_A^2}, \tag{C-25}$$

where

$$u_A = \sqrt{u_{cal}^2 + u_{UUT,b}^2} \,. \tag{C-26}$$

Substituting Eqs. (C-23), (C-24) and (C-25) in Eq. (C-21) yields, after a little algebra,

$$f(e_{UUT,b} \mid \delta) = \frac{1}{\sqrt{2\pi}u_\beta} e^{-(e_{UUT,b} - \beta)^2/2u_\beta^2}, \tag{C-27}$$

where

$$\beta = \frac{u_{UUT,b}^2}{u_A^2} \delta, \tag{C-28}$$

and

$$u_\beta = \frac{u_{UUT,b} u_{cal}}{u_A} \,. \tag{C-29}$$

C.2.1 *CFAR* Revisited

The UUT attribute is in-tolerance if $e_{UUT,b}$ lies within its tolerance limits and is accepted without correction as in-tolerance if δ falls within acceptable limits, sometimes referred to as "guardband limits."[55] Following the notation of [2], we denote the tolerance limits for $e_{UUT,b}$ as $-L_1$ and L_2 and the acceptance limits for δ as $-A_1$ and A_2. Then the UUT attribute is in-tolerance if $-L_1 \leq e_{UUT,b} \leq L_2$ and is observed in-tolerance if $-A_1 \leq \delta \leq A_2$.

While the use of acceptance limits that differ from tolerance limits are relevant to expressions for computing *UFAR*, *CFAR*, *FRR* and *CFRR*, given earlier, they are not relevant to computing these metrics using Bayesian analysis. What we want to determine using Bayes' theorem is simply the probability that the UUT attribute is out-of-tolerance, given a calibration result δ. This can be expressed as

$$p(out \mid \delta) = 1 - p(in \mid \delta), \tag{C-30}$$

where

---

[55] The term "guardband" was used in the original and in early publications on the subject [C-4, C-5]. However, the alternative spelling "guard band" has found recent usage.

$$p(in \mid \mathcal{E}) = \int_{-L_1}^{L_2} f(e_{UUT,b} \mid \delta) de_{UUT,b} \, . \tag{C-31}$$

If this result is not satisfactory, an adjustment or other correction of the UUT attribute value can be made, i.e., there is no need for special adjustment limits.

If $e_{UUT,b}$ and δ are normally distributed, as in Eqs. (C-23) and (C-24) then we can use Eq. (C-27) to get

$$
\begin{aligned}
p(in \mid \delta) &= \frac{1}{\sqrt{2\pi} u_\beta} \int_{-L_1}^{L_2} e^{-(e_{UUT,b} - \beta)^2 / 2u_\beta^2} de_{UUT,b} \\
&= \frac{1}{\sqrt{2\pi}} \int_{-(L_1+\beta)/u_\beta}^{(L_2-\beta)/u_\beta} e^{-\zeta^2/2} d\zeta \\
&= \Phi\left(\frac{L_2 - \beta}{u_\beta}\right) + \Phi\left(\frac{L_1 + \beta}{u_\beta}\right) - 1,
\end{aligned}
\tag{C-32}
$$

where Φ is the normal probability distribution function.

As an example, consider a case where $L_1 = L_2 = 10$ mV, $p(in) = 0.85$ and $u_{cal} = 1.2755$ mV. With ±10 mV tolerance limits and 0.85 in-tolerance probability, we have [2][56]

$$
\begin{aligned}
u_{UUT,b} &= \frac{L_2}{\Phi^{-1}\left(\dfrac{1 + p(in)}{2}\right)} \\
&= \frac{10 \text{ mV}}{\Phi^{-1}\left(\dfrac{1 + 0.85}{2}\right)} = 6.9467 \text{ mV} \, ,
\end{aligned}
$$

and Eqs. (C-26) and (C-28) – (C-29) yield

$$
\begin{aligned}
u_A &= \sqrt{(6.9467)^2 + (1.2755)^2} \text{ mV} \, , \\
&= 7.0628 \text{ mV}, \\
\beta &= \frac{(6.9467)^2}{(7.0628)^2} \delta \\
&= 0.9674 \delta,
\end{aligned}
$$

and

$$
\begin{aligned}
u_\beta &= \frac{(6.9467)(1.2755)}{7.0628} \text{ mV} \, . \\
&= 1.2545 \text{ mV}.
\end{aligned}
$$

---

[56] The function Φ^{-1} is the inverse normal distribution function found in most spreadsheet applications.

Suppose we obtain a calibration result of $\delta = 8$ mV, giving a value $\beta = (0.9674)(8 \text{ mV}) = 7.7395$ mV. With this value inserted in Eq. (C-32), along with the above value for u_β, we have

$$p(in \mid \delta) = \Phi\left(\frac{10 - 7.7395}{1.2545}\right) + \Phi\left(\frac{10 + 7.7395}{1.2545}\right) - 1,$$

$$= \Phi(1.8019) + \Phi(14.1407) - 1$$

$$\cong \Phi(1.8019) = 0.9642.$$

which corresponds to $p(out \mid \delta) = 1 - p(in \mid \delta) = 0.0358$.[57] Note that a calibration result of $\delta = 8$ mV is within the tolerance limits of ± 10 mV. The usual practice in calibration would be to pronounce the UUT attribute in-tolerance, based on this result and an adjustment or other correction would be made.

If the attribute is passed with no adjustment or other correction, we have about a 3.6 percent chance of having falsely accepted an out-of-tolerance attribute. Whether this is acceptable, depends on requirements for compliance with predetermined criteria sometimes based on cost, risk or other factors.

Suppose we had obtained a calibration result of $\delta = 5.0$ mV. Then performing the calculations gives $p(out \mid \delta) \cong 0.0019$ or a *CFAR* of about 0.2 percent. If we had $\delta = 9.0$ mV, we get $p(out \mid \delta) \cong 0.1513$ or a *CFAR* of more than 15 percent.[58]

C.2.2 *CFAR* After Adjustment

If an adjustment or other correction is made to reduce δ to zero, the relevant pdf is

$$p(in \mid 0) = \Phi\left(\frac{L_2}{u_\beta}\right) + \Phi\left(\frac{L_1}{u_\beta}\right) - 1.$$

In this expression, note that, although δ has been reduced to zero, the uncertainty u_{cal} is still present, and must remain included in u_β.[59]

In the present example (assuming no change in u_β), we get

---

[57] The values of the normal distribution functions were obtained using the Microsoft Excel workbook function NORMSDIST. The computations of $p(out \mid \delta)$ were made using MS Excel and verified with ISG's RiskGuard freeware [C-3].

[58] An ill-advised practice has emerged in recent years to set acceptance limits by reducing the tolerance limits by 2 times the calibration process uncertainty. In the examples presented here, we would have $A_1 = A_2 = L_2 - 2u_{cal} = 10 - (2)(1.2755 \text{ mV}) = 7.449$ mV. Using these limits would trigger adjustments in all cases shown except the case where $\delta = 5$ mV. Modifications of this theme can be found in [C-6].

[59] In some cases, making a physical adjustment introduces error in addition to the pre-adjustment calibration error. Since u_{cal} is the uncertainty in the total calibration error ε_{cal}, it may need to be modified to include the uncertainty in the error due to adjustment.

$$p(in \mid 0) = 2\Phi\left(\frac{L_2}{u_\beta}\right) - 1$$

$$= 2\Phi\left(\frac{10}{1.2545}\right) - 1$$

$$\cong 1.0000,$$

and a false accept risk $p(out \mid C) \cong 0$. For this example, even if adjustment change ε_{cal} in some unknown way that causes u_β to be doubled, we would still get

$$p(in \mid 0) = 2\Phi\left(\frac{5}{1.2545}\right) - 1 = 0.99993$$

$$\cong 1.0000.$$

C.2.3   Bench-Level Implementation of the Method

If normal distributions can be assumed for $f(\delta \mid e_{UUT,b})$ and $f(e_{UUT,b})$, then Bayes' theorem can be applied to compute the probability of accepting an out-of-tolerance UUT attribute $p(out \mid \delta)$. This is evident from the fact that the required quantities, shown in Eq. (C-26), Eqs. (C-28) – (C-29) and Eq. (C-32), can all be computed using a spreadsheet application or other program that can be made available to calibrating personnel. No numerical integrations or other iterative routines are required. In short, a decision to take corrective action can be made at the calibration bench in response to simple data entry involving a few keystrokes. For this reason, the method has been called a "bench-level" method [C-2].

Appendix C References

[C-1]   Castrup, H. and Castrup, S., "Calibration Scenarios for Uncertainty Analysis," Submitted for inclusion in *Proc. NCSLI Workshop & Symposium*, August 2008, Orlando.

[C-2]   Castrup, H., "Risk Analysis Methods for Complying with Z540.3," *Proc. NCSLI Workshop & Symposium*, July 2007, St. Paul.

[C-3]   ISG, *RiskGuard 2.0*, © 2003-2007, Integrated Sciences Group, www.isgmax.com.

[C-4]   Hutchinson, B., "Setting Guardband Test Limits to Satisfy MIL-STD-45662A Requirements," *Proc. NCSL Workshop & Symposium*, August 1991, Albuquerque.

[C-5]   Deaver, D., "Guardbanding with Confidence," *Proc. NCSL Workshop & Symposium*, August 1994, Chicago.

[C-6]   ASME, B89.7.3.1-2001, *Guidelines for Decision Rules: Considering Measurement Uncertainty in Determining Conformance to Specifications*, The American Society of Mechanical Engineers, March 18, 2002.

Appendix D: Derivation of Key Bayesian Expressions

D.1 Introduction

Bayesian risk analysis estimates in-tolerance probabilities and attribute biases for both a unit under test (UUT) and a set of independent measuring and test instruments (MTE). The estimation of these quantities is based on measurements of a UUT attribute value made by the MTE set and on certain information regarding UUT and MTE attribute uncertainties. The method accommodates arbitrary test uncertainty ratios between MTE and UUT attributes and applies to MTE sets comprised of any number of instruments [D-1].

To minimize abstraction of the discussion, the treatment in this appendix focuses on restricted cases in which both MTE and UUT attribute values are normally distributed and are maintained within two-sided symmetric tolerance limits. This should serve to make the mathematics more condensed. Despite these mathematical restrictions, the methodological framework is entirely general. Extension to cases involving one-sided tolerances and asymmetric attribute distributions merely calls for more mathematical brute force.

A Comment on Nomenclature

The nomenclature used in this appendix is that used in the references from which it has been previously reported [27, 30]. That this notation differs from the notation in other parts of this Annex is acknowledged. The reason for this departure is that the notation used elsewhere yields expressions that are typographically awkward. Since this appendix is essentially self-contained, it is hoped that the change of notation will not hamper the ability of readers to be able to follow the treatment.

D.2 Computation of In-Tolerance Probabilities

D.2.1 UUT In-Tolerance Probability

Whether a UUT provides a stimulus, indicates a value, or exhibits an inherent property, the declared value of its output, indicated value, or inherent property, is said to reflect some underlying "true" value. A frequency reference is an example of a stimulus, a frequency meter reading is an example of an indicated value, and a gage block dimension is an example of an inherent property. Suppose for example that the UUT is a voltmeter measuring a (true) voltage of 10.01 mV. The UUT meter reading (10.00 mV or 9.99 mV, or some such) is the UUT's "declared" value. As another example, consider a 5 cm gage block. The declared value is 5 cm. The unknown true value (gage-block dimension) may be 5.002 cm, or 4.989 cm, or some other value.

The UUT declared value is assumed to deviate from the true value by an unknown amount. Let Y_0 represent the UUT attribute's declared value and define a random variable ε_0 as the deviation of Y_0 from the true value. The variable ε_0 is assumed a priori to be normally distributed with zero mean and standard deviation σ_0. The tolerance limits for ε_0 are labeled $\pm L_0$, i.e., the UUT is considered in-tolerance if $-L_0 \leq \varepsilon_0 \leq L_0$.

A set of n independent measurements are also taken of the true value using n MTE. Let Y_i be the declared value representing the ith MTE's measurement. The observed differences between UUT and MTE declared values are labeled according to

$$X_i \equiv Y_0 - Y_i, \quad i = 1, 2, \text{L}, n \tag{D-1}$$

The quantities X_i are assumed to be normally distributed random variables with mean ε_0 standard deviation σ_i.

Designating the tolerance limits of the ith MTE attribute by $\pm L_i$, the ith MTE is considered in-tolerance if $\varepsilon_0 - L_i \leq X_i \leq \varepsilon_0 + L_i$. Populations of MTE measurements are not expected to be systematically biased. This is the usual assumption made when MTE are chosen either randomly from populations of like instruments or when no foreknowledge of MTE bias is available. Individual unknown MTE biases *are* assumed to exist. Accounting for this bias is done by treating individual instrument bias as a random variable and estimating its variance. Estimating this variance is the subject of Section D3. Estimating biases is covered in Section D6.

In applying Bayesian methodology, we work with a set of variables r_i, called *dynamic accuracy ratios* (or dynamic inverse uncertainty ratios) defined according to

$$r_i \equiv \frac{\sigma_0}{\sigma_i}, \quad i = 1, 2, \text{L}, n \tag{D-2}$$

The adjective "dynamic" distinguishes these accuracy ratios from their usual static or "nominal" counterparts, defined by L_0 / L_i, $i = 1, 2, \cdots, n$. The use of the word "dynamic" underscores the fact that each r_i defined by Eq. (D-2) is a quantity that changes as a function of time passed since the last calibrations of the UUT and of the ith MTE. This dynamic character exists because generally both UUT and MTE population standard deviations (bias uncertainties) grow with time since calibration.

Let P_0 be the probability that the UUT is in-tolerance at some given time since calibration. Using these definitions, we can write

$$P_0 = \Phi(a_+) + \Phi(a_-) - 1, \tag{D-3}$$

where Φ is the normal distribution function defined by

$$\Phi(a_\pm) = \frac{1}{\sqrt{2\pi}} \int_{-\infty}^{a_\pm} e^{-\zeta^2/2} d\zeta, \tag{D-4}$$

and where

$$a_\pm = \frac{\sqrt{1 + \sum r_i^2}\left(L_0 \pm \frac{\sum X_i r_i^2}{1 + \sum r_i^2} \right)}{\sigma_0}. \tag{D-5}$$

In these expressions and in others to follow, all summations are taken over $i = 1, 2, \cdots, n$. The derivation of Eqs. (D-3) and (D-5) is presented in Section D.5. Note that the time dependence of P_0 is in the time dependence of a_+ and a_-. The time dependence of a_+ and a_- is, in turn, in the time dependence of r_i.

D.2.2   MTE In-Tolerance Probability

Just as the random variables $X_1, X_2, \cdots, X_n$ are MTE-measured deviations from the UUT declared value, they are also UUT-measured deviations from MTE declared values. Therefore, it is easy to see that by reversing its role, the UUT can act as a MTE. In other words, *any* of the n MTE can be regarded as the UUT, with the original UUT performing the service of a MTE. For example, focus on the ith (arbitrarily labeled) MTE and swap its role with that of the UUT. This results in the following transformations:

$$X_1' = X_1 - X_i$$
$$X_2' = X_2 - X_i$$
$$\text{M}$$
$$X_i' = -X_i$$
$$\text{M}$$
$$X_n' = X_n - X_i \, ,$$

where the primes indicate a redefined set of measurement results. Using the primed quantities, the in-tolerance probability for the ith MTE can be determined just as the in-tolerance probability for the UUT was determined earlier. The process begins with calculating a new set of dynamic accuracy ratios. First, we set

$$\sigma_0' = \sigma_i, \quad \sigma_1' = \sigma_1, \quad \sigma_2' = \sigma_2, \cdots, \sigma_i' = \sigma_0, \cdots, \sigma_n' = \sigma_n \, .$$

Given these label reassignments, the needed set of accuracy ratios can be obtained using Eq. (D-2), i.e.,

$$r_i' = \sigma_0' / \sigma_i', \quad i = 1,2,\text{L}\ ,n \, .$$

Finally, the tolerance limits are relabeled for the UUT and the ith MTE according to $L_0' = L_i$ and $L_i' = L_0$.

If we designate the in-tolerance probability for the ith MTE by P_i and we substitute the primed quantities obtained above, Eqs. (D-3) and (D-5) become

$$P_i = \Phi(a_+') + \Phi(a_-') - 1 \, ,$$

and

$$a_\pm' = \frac{\sqrt{1 + \sum r_i'^2}\left(L_0' \pm \dfrac{\sum X_i' r_i'^2}{1 + \sum r_i'^2} \right)}{\sigma_0'} \, .$$

Applying similar transformations yields in-tolerance probabilities for the remaining MTE.

D.3   Computation of Variances
D.3.1   Variance in Instrument Bias

Computing the uncertainties in UUT and MTE attribute biases involves establishing the relationship between attribute uncertainty growth and time since calibration. Several models have been used to describe this relationship [D-2].

To illustrate the computation of bias uncertainties, the simple negative exponential model will be used here. With the exponential model, if t represents the time elapsed since calibration, then the corresponding UUT in-tolerance probability $R(t)$ is given by

$$R_0(t) = R_0(0) e^{-\lambda_0 t},$$
(D-6)

where the attribute λ_0 is the out-of-tolerance rate for the UUT in question, and $R_0(0)$ is the in-tolerance probability immediately following calibration. Note that setting $R_0(0) < 1$ acknowledges that a finite measurement uncertainty exists immediately following calibration. The attributes λ and $R_0(0)$ are usually obtained from analysis of a homogeneous population of instruments of a given model number or type [D-2].

With the exponential model, for a given end-of-period in-tolerance target, $R_0{}^*$, the attributes λ and $R_0(0)$ determine the calibration interval T_0 for a population of UUT attributes according to

$$T_0 = -\frac{1}{\lambda_0} \ln\left[\frac{R_0^*}{R_0(0)}\right].$$
(D-7)

For a UUT attribute whose acceptable values are bounded within tolerance limits $\pm L_0$, the in-tolerance probability can also be written, assuming a normal distribution, as

$$R_0(t) = \frac{1}{\sqrt{2\pi\sigma_b^2(t)}} \int_{-L_0}^{L_0} e^{-\zeta^2/2c_b^2} d\zeta,$$
(D-8)

where $\sigma_b(t)$ is the expected standard deviation of the attribute bias at time t. Substituting Eq. (D-8) in Eq. (D-6) gives

$$\sigma_b(t) = \frac{L_0}{\Phi^{-1}\left(\frac{1 + R_0(0)e^{-\lambda_0 t}}{2}\right)},$$
(D-9)

where Φ^{-1} is the inverse of the normal distribution function. Substituting from Eq. (D-7) yields the UUT attribute bias standard deviation at the time of test or calibration

$$\sigma_b(T_0) = \frac{L_0}{\Phi^{-1}\left(\frac{1 + R_0^*}{2}\right)},$$
(D-10)

Let t_i be the time elapsed since calibration of the ith MTE at the time of the UUT calibration. Then, if the exponential model is applicable for MTE in-tolerance probabilities, using L_i, t_i, and $R_i(0)$ in Eq.(D-9) in place of L_0, t, and $R_0(0)$ yields the appropriate MTE bias standard deviations.

D.3.2   Treatment of Multiple Measurements

In previous discussions, the quantities X_i are treated as single measurements of the difference between the UUT attribute and the ith MTE's measurement. Yet, in most applications, testing or calibration of workload attributes is not limited to single measurements. Instead, multiple measurements are usually taken. Instead of n individual measurements, we will ordinarily be dealing with n sets or *samples* of measurements. In these samples, let n_i be the number of measurements taken using the ith MTE's attribute, and let

$$X_{ij} = Y_0 - Y_{ij}$$

be the jth of these measurements. The sample mean and standard deviation are given in the usual way:

$$X_i = \frac{1}{n_i} \sum_{j=1}^{n_i} X_{ij} \qquad \text{(D-11)}$$

and

$$s_i^2 = \frac{1}{n_i - 1} \sum_{j=1}^{n_i} \left(X_{ij} - X_i \right)^2 . \qquad \text{(D-12)}$$

The variance associated with the mean of measurements made using the ith MTE's attribute is given by

$$\sigma_i^2 = \sigma_{b_i}^2 + s_i^2 / n_i + \sigma_{i,other}^2 , \qquad \text{(D-13)}$$

where the variables σ_{bi} is the bias uncertainty of the ith MTE and $\sigma_{i,other}$ is the uncertainty due to other test or calibration error sources. The square root of this variance will determine the quantities r_i defined in Eq. (D-2).

Note that including sample variances is restricted to the estimation of MTE attribute variances. UUT attribute variance estimates contain only the terms σ_{bi} and $\sigma_{\delta i}$. This underscores what is sought in constructing the pdf $f(\varepsilon_0 \mid \mathbf{X})$. What we seek are estimates of the in-tolerance probability and bias of the UUT attribute. In this, we are interested in the attribute as an entity distinct from process uncertainties involved in its measurement.

D.4. Example

To illustrate the Bayesian method, consider the following question that arose during a proficiency audit conducted on board the USS Frank Cable [D-3]:

> "We have three instruments with identical tolerances of ±10 psi. One instrument measures an unknown quantity as 0 psi, the second measures +6 psi, and the third measures +15 psi. According to the first instrument, the third one is out-of-tolerance. According to the third instrument, the first one is out-of-tolerance. Which is out-of-tolerance?"

Of course, it is never possible to say with certainty whether a given instrument is in- or out-of-tolerance. Instead, the best we can do is to try to evaluate out-of-tolerance or in-tolerance probabilities. The application of the method to the proficiency audit example follows.

The measurement configuration is shown in Figure D-1 and tabulated in column 1 of Table D-1. For discussion purposes, let instrument 1 act the role of a UUT and label it's indicated or "declared" value as Y_0. Likewise, let instruments 2 and 3 function as MTE, label their declared values as Y_1 and Y_2, respectively, (the "1" and "2" subscripts label MTE$_1$ and MTE$_2$) and define the variables.

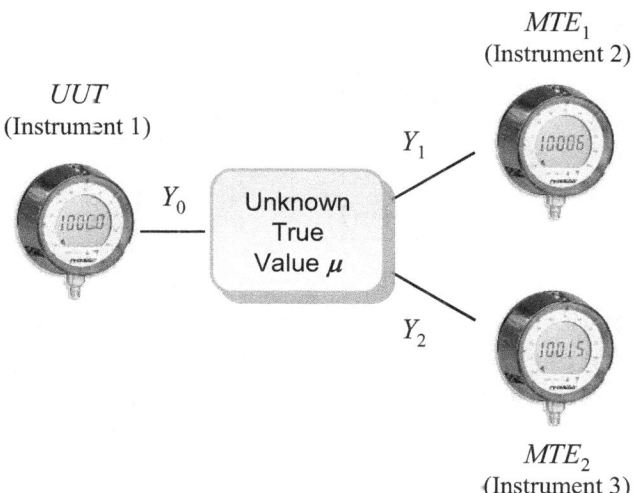

Figure D-1. Proficiency Audit Example

Three instruments measure an unknown value. This value may be external to all three instruments or generated by one or more of them. Instrument 1 is arbitrarily labeled the UUT. Instruments 2 and 3 are employed as MTE. The tolerance limits for each of the instruments are ±10 psi.

Table D-1. Proficiency Audit Results Arranged for Bayesian Analysis.

The "0" subscript labels the UUT.

UUT = MTE₁	UUT = MTE₂	UUT = MTE₃
$L_0 = 10$	$L_0' = 10$	$L_0'' = 10$
$L_1 = 10$	$L_1' = 10$	$L_1'' = 10$
$L_2 = 10$	$L_2' = 10$	$L_2'' = 10$
$Y_0 = 0$	$Y_0' = 6$	$Y_0'' = 15$
$Y_1 = 6$	$Y_1' = 0$	$Y_1'' = 6$
$Y_2 = 15$	$Y_2' = 6$	$Y_2'' = 0$
$X_1 = -6$	$X_1' = 6$	$X_1'' = 9$
$X_2 = -15$	$X_2' = -9$	$X_2'' = 15$

With these designations, we have $Y_0 = 0$, $Y_1 = 6$, and $Y_2 = 15$. Thus,

$$X_1 = Y_0 - Y_1 = -6$$
$$X_2 = Y_0 - Y_2 = -15.$$

Since it was not stated otherwise, we assume that the in-tolerance probabilities for all three instruments are equal. The three instruments are managed to the same R^* target, have the same tolerances, and are calibrated in the same way using the same equipment and procedures. Therefore, their standard deviations when the measurements were made should be about equal. According to Eq. (D-2), the dynamic accuracy ratios are then

$$r_1 = r_2 = 1.$$
$$r_1 = r_2 = 1.$$

Then, by using Eq. (D-5), we get

$$a_{\pm} = \frac{\sqrt{1+(1+1)}\left[10 \pm \dfrac{-6-15}{1+(1+1)}\right]}{\sigma_0}$$

$$= \frac{\sqrt{3}(10\,\mathrm{m}7)}{\sigma_0}.$$

Calculation of the standard deviation σ_0 calls for some supplemental information. The quantity σ_0 is an *a priori* estimate of the bias standard deviation for the UUT attribute value of interest. In making such estimates, it is usually assumed that the UUT is drawn at random from a population. If knowledge of the population's uncertainty is available, then an estimate for σ_0 can be obtained.

For the instruments used in the proficiency audit, it was determined that the population uncertainty is managed to achieve an in-tolerance probability of $R^* = 0.72$ at the end of the calibration interval and that $R(0) \cong 1$ for each instrument. As a fair approximation, we assume that we can use average-over-period in-tolerance probabilities for each $R(t)$ in this example. With the exponential model, if $R(0) = 1$, the average in-tolerance probability is equal to the square root of the reliability target R^*. Using this observation in Eq. (D-10) yields

$$\sigma_0(t) = \frac{10\ \mathrm{psi}}{\Phi^{-1}\left(\dfrac{1+\sqrt{0.72}}{2}\right)}$$

$$= 6.97\ \mathrm{psi}.$$

Substituting in the expression for $a_{\pm}$ above gives

$$a_{\pm} = \frac{\sqrt{3}(10\,\mathrm{m}7)}{6.97}$$

$$= 2.49 \pm 1.74.$$

Thus, the in-tolerance probability for the UUT (instrument 1) is

$$P_0 = \Phi(0.75) + \Phi(4.23) - 1$$

$$= 0.77 + 1.00 - 1$$

$$= 0.77.$$

To compute the in-tolerance probability for MTE$_1$ (instrument 2), the UUT and MTE$_1$ swap roles. By using the transformations of Table D-1, we have

$$X_1' = -X_1$$

$$= 6$$

$$X_2' = X_2 - X_1$$

$$= -9$$

in place of X_1 and X_2 in Eq. (D-5). Recalling that $\sigma_0' = \sigma_0$ in this example gives

$$a_\pm' = \frac{\sqrt{1+(1+1)}\left[10 \pm \dfrac{6-9}{1+(1+1)}\right]}{\sigma_0'}$$

$$= \frac{\sqrt{3}\,(10 \pm 1)}{6.97}$$

$$= 2.49 \pm 0.25.$$

Thus, by Eq. (D-3), the in-tolerance probability for MTE 1 (instrument 2) is

$$P_1 = \Phi(2.24) + \Phi(2.73) - 1$$

$$= 0.99 + 1.00 - 1$$

$$= 0.99.$$

In computing the in-tolerance probability for MTE$_2$, the UUT and MTE$_2$ swap roles. Thus

$$X_1' = X_1 - X_2$$

$$= 15$$

$$X_2' = -X_2$$

$$= 9.$$

Using these quantities in Eq. (D-5) and setting $\sigma_0' = \sigma_0$ gives

$$a_\pm' = 2.49 \pm 1.99.$$

Thus, by Eq. (D-3), the in-tolerance probability for MTE 2 (instrument 3) is

$$P_2 = \Phi(4.47) + \Phi(0.50) - 1$$

$$= 1.00 + 0.69 - 1$$

$$= 0.69.$$

Summarizing these results, we estimate a roughly 77% in-tolerance probability for instrument 1, a 99% in-tolerance probability for instrument 2, and a 69% in-tolerance probability for instrument 3. As shown earlier, the instruments in the proficiency audit example are managed to an end-of-period in-tolerance probability of 0.72. They are candidates for calibration if their intolerance probabilities fall below 72%. Therefore, instrument 3 should be calibrated.

D.5   Derivation of Eq. (D-3)

Let the vector $\mathbf{X}$ represent the random variables $X_1, X_2, \cdots, X_n$ obtained from n independent MTE measurements of ε_0. We seek the conditional pdf for ε_0, given $\mathbf{X}$, that will, when integrated over $[-L_0, L_0]$, yield the conditional probability P_0 that the UUT is in-tolerance. This pdf will be represented by the function $f(\varepsilon_0 \mid \mathbf{X})$. From basic probability theory, we have[60]

---

[60] See Section 4.3.4.

$$f\left(\varepsilon_0 \mid \mathbf{X}\right) = \frac{f\left(\mathbf{X} \mid \varepsilon_0\right) f\left(\varepsilon_0\right)}{f\left(\mathbf{X}\right)}, \tag{D-14}$$

where

$$f\left(\varepsilon_0\right) = \frac{1}{\sqrt{2\pi}\sigma_0} e^{-\varepsilon_0^2/2\sigma_0^2}. \tag{D-15}$$

In Eq. (D-14), the pdf $f(\mathbf{X} \mid \varepsilon_0)$ is the probability density for observing the set of measurements $X_1, X_2, \cdots, X_n$, given that the bias of the UUT is ε_0. The pdf $f(\varepsilon_0)$ is the probability density for UUT biases.

Since the components of $\mathbf{X}$ are s-independent, we can write

$$f\left(\mathbf{X} \mid \varepsilon_0\right) = f\left(X_1 \mid \varepsilon_0\right) f\left(X_2 \mid \varepsilon_0\right) \cdots f\left(X_n \mid \varepsilon_0\right), \tag{D-16}$$

where

$$f\left(X_i \mid \varepsilon_0\right) = \frac{1}{\sqrt{2\pi}\sigma_i} e^{-\left(X_i - \varepsilon_0\right)^2/2\sigma_i^2}, \quad i = 1, 2, \cdots, n. \tag{D-17}$$

Note that Eq. (D-17) states, for the present discussion, we assume the measurements of ε_0 to be normally distributed with a population mean value ε_0 (the UUT "true" value) and a standard deviation σ_i. At this point, we do not provide for an unknown bias in the ith MTE.[61] As we will see, the Bayesian methodology will be used to estimate this bias, based on the results of measurement and on estimated measurement uncertainties.

Combining Eqs. (D-14) through (D-17) gives

$$\begin{aligned}
f\left(\mathbf{X} \mid \varepsilon_0\right) f\left(\varepsilon_0\right) &= C\exp\left\{-\frac{1}{2}\left[\frac{\varepsilon_0^2}{\sigma_0^2} + \sum_{i=1}^{n} \frac{\left(X_i - \varepsilon_0\right)^2}{\sigma_i^2}\right]\right\} \\
&= C\exp\left\{-\frac{1}{2\sigma_0^2}\left[\varepsilon_0^2 + \sum_{i=1}^{n} r_i^2\left(X_i - \varepsilon_0\right)^2\right]\right\} \\
&= Ce^{-G(\mathbf{X})}\exp\left\{-\frac{1}{2\sigma_0^2}\left[\left(1 + \sum r_i^2\right)\left(\varepsilon_0 - \frac{\sum X_i r_i^2}{1 + \sum r_i^2}\right)^2\right]\right\},
\end{aligned} \tag{D-18}$$

where C is a normalization constant. The function $G(\mathbf{X})$ contains no ε_0 dependence and its explicit form is not of interest in this discussion.

The pdf $f(\mathbf{X})$ is obtained by integrating Eq. (D-18) over all values of ε_0. To simplify the notation, we define

---

[61] It can be readily shown that, if the bias of a MTE is unknown, the best estimate for the *population* of its measurements is the true value being measured, i.e., zero bias. This is an important *a priori* assumption in applying the SMPC methodology.

$$\alpha = \sqrt{1 + \sum r_i^2} \tag{D-19}$$

and

$$\beta = \frac{\sum X_i r_i^2}{1 + \sum r_i^2}. \tag{D-20}$$

Using Eqs. (D-19) and (D-20) in Eq. (D-18) and integrating over ε_0 gives

$$\begin{aligned} f(\mathbf{X}) &= \int_{-\infty}^{\infty} f(\mathbf{X} \mid \varepsilon_0) f(\varepsilon_0) d\varepsilon_0 \\ &= C e^{-G(\mathbf{X})} \int_{-\infty}^{\infty} e^{-\alpha^2 (\varepsilon_0 - \beta)^2 / 2\sigma_0^2} d\varepsilon_0 \\ &= C e^{-G(\mathbf{X})} \frac{\sqrt{2\pi}\sigma_0}{\alpha}. \end{aligned} \tag{D-21}$$

Dividing Eq. (D-21) into Eq. (D-18) and substituting from Eq. (D-14) yields the pdf

$$f(\varepsilon_0 \mid \mathbf{X}) = \frac{1}{\sqrt{2\pi}\,(\sigma_0 / \alpha)} e^{-(\varepsilon_0 - \beta)^2 / 2(\sigma_0 / \alpha)^2}. \tag{D-22}$$

As we can see, ε_0 conditional on $\mathbf{X}$ is normally distributed with mean β and standard deviation σ_0 / α. The in-tolerance probability for the UUT is obtained by integrating Eq. (D-22) over $[-L_0, L_0]$. With the aid of Eq. (D-5), this results in

$$\begin{aligned} P_0 &= \frac{1}{\sqrt{2\pi}\,(\sigma_0 / \alpha)} \int_{-L_0}^{L_0} e^{-(\varepsilon_0 - \beta)^2 / 2(\sigma_0 / \alpha)^2} d\varepsilon_0 \\ &= \frac{1}{\sqrt{2\pi}} \int_{-(L_0 + \beta)/(\sigma_0/\alpha)}^{(L_0 - \beta)/(\sigma_0/\alpha)} e^{-\zeta^2 / 2} d\zeta \\ &= \Phi(a_-) - \Phi(a_+) \\ &= \Phi(a_+) + \Phi(a_-) - 1, \end{aligned}$$

which is Eq. (D-3) with α and β as defined in Eqs. (D-19) and D.21).

D.6 Estimation of Biases

Obtaining the conditional pdf $f(\varepsilon_0 \mid \mathbf{X})$ allows us to compute moments of the UUT attribute distribution. Of particular interest is the first moment or *distribution mean*. The UUT distribution mean is the conditional expectation value for the bias ε_0. Thus, the UUT attribute bias is estimated by

$$\begin{aligned} \beta_0 &= E(\varepsilon_0 \mid \mathbf{X}) \\ &= \int_{-\infty}^{\infty} \varepsilon_0 f(\varepsilon_0 \mid \mathbf{X}) d\varepsilon_0. \end{aligned} \tag{D-23}$$

Substituting from Eq. (D-22) and using Eq. (D-20) gives

$$\beta_0 = \frac{\sum X_i r_i^2}{1 + \sum r_i^2} \qquad \text{(D-24)}$$

Similarly, bias estimates can be obtained for the MTE set by making the transformations described in Section D.2.2; for example, the bias of MTE 1 is given by

$$\beta_1 = E(\varepsilon_1 \mid \mathbf{X}') = \frac{\sum X_i' r_i'^2}{1 + \sum r_i'^2}. \qquad \text{(D-25)}$$

To exemplify bias estimation, we again turn to the proficiency audit question. By using Eqs. (D-24) and (D-25), and by recalling that $\sigma_0 = \sigma_1 = \sigma_2$, we get

$$\text{Instrument 1 (UUT) bias:} \quad \beta_0 = \frac{-6 - 15}{1 + (1 + 1)} = -7$$

$$\text{Instrument 2 (MTE 1) bias:} \quad \beta_1 = \frac{6 - 9}{1 + (1 + 1)} = -1$$

$$\text{Instrument 3 (MTE 2) bias:} \quad \beta_2 = \frac{15 + 9}{1 + (1 + 1)} = 8$$

If desired, these bias estimates could serve as correction factors for the three instruments. If used in this way, the quantity 7 would be added to all measurements made with instrument 1. The quantity 1 would be added to all measurements made with instrument 2. And, the quantity 8 would be subtracted from all measurements made with instrument 3.[62]

Note that all biases are within the stated tolerance limits (± 10) of the instruments, which might encourage users to continue to operate their instruments with confidence. However, recall that the in-tolerance probabilities computed in Section D.4 showed only a 77% chance that instrument 1 was in-tolerance and an even lower 69% chance that instrument 3 was in-tolerance. Such results tend to provide valuable information from which to make cogent judgments regarding instrument calibration.

D.7   Bias Confidence Limits

Another variable that can be useful in making decisions based on measurement results is the range of the confidence limits for the estimated biases. Estimating confidence limits for the computed biases β_0 and β_i, $i = 1, 2, \cdots, n$, means first determining the statistical probability density functions for these biases. From Eq. (D-24) we can write

---

[62] Since all three instruments are considered *a priori* to be of equal accuracy, the best estimate of the true value of the measured quantity would be the average of the three measured deviations: $\varepsilon_0 = (0 + 6 + 15)/3 = 7$. Thus, a zero reading would be indicative of a bias of -7, a $+6$ reading would be indicative of a bias of -1, and a $+15$ reading would be indicative of a bias of $+8$. These are the same estimates we obtained with SMPC. Obviously, this is a trivial example. Things become more interesting when each measurement has a different uncertainty, i.e., when $\sigma_0 \neq \sigma_1 \neq \sigma_2$.

$$\beta_0 = \sum_{i=1}^{n} c_i X_i, \tag{D-26}$$

where

$$c_i = \frac{r_i^2}{1 + \sum r_i^2}. \tag{D-27}$$

With this convention, the probability density function of β_0 can be written:

$$\begin{aligned} f(\beta_0) &= f(\sum c_i X_i) \\ &= f(\sum \psi_i), \end{aligned} \tag{D-28}$$

where

$$\psi_i = c_i X_i. \tag{D-29}$$

Although the coefficients c_i, $i = 1, 2, \cdots, n$, are in the strictest sense random variables, to a first approximation, they can be considered fixed coefficients of the variables X_i. Since the variables X_i are normally distributed (see Eq. (D-17)), the variables ψ_i are also normally distributed. The appropriate expression is

$$f(\psi_i) = \frac{1}{\sqrt{2\pi}\sigma_{\psi_i}} e^{-(\psi_i - \eta_i)^2 / 2c_{\psi_i}^2}, \tag{D-30}$$

where

$$\sigma_{\psi_i} = c_i \sigma_i \tag{D-32}$$

and

$$\eta_i = c_i \varepsilon_0. \tag{D-32}$$

Since the variables ψ_i are normally distributed, their linear sum is also normally distributed:

$$\begin{aligned} f(\sum \psi_i) &= \frac{1}{\sqrt{2\pi}\sigma} e^{-(\sum \psi_i - \eta)^2 / 2\sigma^2} \\ &= \frac{1}{\sqrt{2\pi}\sigma} e^{-(\beta_0 - \eta)^2 / 2\sigma^2}, \end{aligned} \tag{D-33}$$

where

$$\sigma = \sqrt{\sum \sigma_{\psi_i}^2}, \tag{D-34}$$

and

$$\eta = \sum \eta_i. \tag{D-35}$$

Equation (D-33) can be used to find the upper and lower confidence limits for β_0. Denoting these limits by β_0^+ and β_0^-, if the desired level of confidence is $p \times 100\%$, then

$$p = \int_{\beta_0^-}^{\beta_0^+} f(\beta_0) d\beta_0,$$

or

$$\int_{\beta_0}^{\infty} f(\beta_0) d\beta_0 = (1-p)/2 = \int_{-\infty}^{\beta_0^-} f(\beta_0) d\beta_0 .$$

Integrating Eq. (D-33) from β_0^+ to ∞ and using Eqs. (D-34) and (D-35) yields

$$1 - \Phi\left(\frac{\beta_0^+ - \eta}{\sigma}\right) = (1-p)/2$$

and

$$\Phi\left(\frac{\beta_0^+ - \eta}{\sigma}\right) = (1+p)/2 .$$

Solving for β_0^+ gives

$$\beta_0^+ = \eta + \sigma \Phi^{-1}\left(\frac{1+p}{2}\right). \tag{D-36}$$

Solving for the lower confidence for β_0^- in the same manner, we begin with

$$\int_{-\infty}^{\beta_0^-} f(\beta_0) d\beta_0 = (1-p)/2 .$$

This yields, with the aid of Eq. (D-23),

$$\Phi\left(\frac{\beta_0^- - \eta}{\sigma}\right) = (1-p)/2 . \tag{D-37}$$

Using the following property of the normal distribution

$$\Phi(-x) = 1 - \Phi(x),$$

we can rewrite Eq. (D-37) as

$$\Phi\left(-\frac{\beta_0^- - \eta}{\sigma}\right) = 1 - (1-p)/2$$

$$= (1+p)/2 ,$$

from whence

$$\beta_0^- = \eta - \sigma \Phi^{-1}\left(\frac{1+p}{2}\right) . \tag{D-38}$$

From Eq. (D-33), the attribute η is seen to be the expectation value for β_0. Our best available estimate for this quantity is the computed UUT bias, namely β_0 itself. We thus write the computed upper and lower confidence limits for β_0 as

$$\beta_0^\pm = \beta_0 \pm \sigma \Phi^{-1}\left(\frac{1+p}{2}\right). \tag{D-39}$$

In like fashion, we can write down the solutions for the MTE biases β_i, $i = 1, 2, \cdots, n$:

$$\beta_i^{\pm} = \beta_i \pm \sigma' \Phi^{-1}\left(\frac{1+p}{2}\right), \quad \text{(D-40)}$$

where

$$\sigma' = \sqrt{\sum c_i'^2 \sigma_i'^2}, \quad \text{(D-41)}$$

and

$$c_i' = \frac{r_i'^2}{1 + \sum r_j'^2}. \quad \text{(D-42)}$$

The variables r_i' in this expression are defined as before.

To illustrate the determination of bias confidence limits, we again turn to the proficiency audit example of Section D.4. In this example,

$$\sigma_0 = \sigma_1 = \sigma_2 = 6.97,$$

and

$$r_1 = r_2 = r_3 = 1.$$

By Eqs. (D-27) and (D-33),

$$c_i = c_i' = \frac{1}{3},$$

and

$$\sigma = \sqrt{\frac{\sigma_1^2}{9} + \frac{\sigma_2^2}{9}}$$
$$= \frac{\sqrt{2}\sigma_0}{3}$$
$$= 3.29 = \sigma'.$$

Substituting in Eqs. (D-39) and (D-40) yields

$$\beta_0^{\pm} = \beta_0 \pm 3.29\,\Phi^{-1}\left(\frac{1+p}{2}\right),$$

$$\beta_1^{\pm} = \beta_1 \pm 3.29\,\Phi^{-1}\left(\frac{1+p}{2}\right),$$

and

$$\beta_2^{\pm} = \beta_2 \pm 3.29\,\Phi^{-1}\left(\frac{1+p}{2}\right).$$

Suppose that the desired confidence level is 95%. Then $p = 0.95$, and

$$\Phi^{-1}\left(\frac{1+p}{2}\right) = \Phi^{-1}(0.975)$$
$$= 1.96,$$

and

$$3.29\,\Phi^{-1}\left(\frac{1+p}{2}\right)=6.4\,.$$

Since $\beta_0 = -7$, $\beta_1 = -1$, and $\beta_2 = +8$, this result, when substituted in the above expressions, gives 95% confidence limits for the estimated biases:

$$-13.4 \le \beta_0 \le -0.6$$
$$-7.4 \le \beta_1 \le 5.4$$
$$1.6 \le \beta_2 \le 14.4\,.$$

Appendix D References

[D-1]   Castrup, H., "Analytical Metrology SPC Methods for ATE Implementation," *Proc. NCSL Workshop & Symposium*, Albuquerque, August 1991.

[D-2]   NCSL, *Establishment and Adjustment of Calibration Intervals*, Recommended Practice RP-1, National Conference of Standards Laboratories, January 1996.

[D-3]   Castrup, H., Intercomparison of Standards: General Case, SAI Comsystems Technical Report, U.S. Navy Contract N00123-83-D-0015, Delivery Order 4M03, March 16, 1984.

Appendix E: True vs. Reported Probabilities

As discussed in Chapter 5, the fact that false accept risk and false reject risk are typically not equal leads to the phenomenon that the observed percent in-tolerance as a result of testing or calibration is typically different from the true percent in-tolerance. It turns out that since, end-of-period reliability targets are nearly always higher than 50%, the perceived or *reported* percent in-tolerance will nearly always be lower than the actual or *true* percent in-tolerance. This was first reported by Ferling in 1984 [E-1] as the "True vs. Reported" problem.

E.1 Appendix E Nomenclature

The variables used in this appendix are described in Table E-1.

Table E-1. Variables Used to Estimate True vs. Reported Percent In-Tolerance.

Variable	Description
$e_{UUT,b}$	the bias of the UUT attribute value at the time of calibration
δ	a measurement result for the UUT attribute bias
$u_{UUT,b}$	the uncertainty in $e_{UUT,b}$, i.e., the standard deviation of the probability distribution of the population of $e_{UUT,b}$ values.[63]
u_{cal}	the standard uncertainty in $u_{UUT,b}$
$-L_1$ and L_2	the tolerance limits for $e_{UUT,b}$
$-A_1$ and A_2	the "acceptance" limits (test limits) for $e_{UUT,b}$
L	the range of values of $e_{UUT,b}$ from $-L_1$ to L_2 (the UUT tolerance limits)
$P(e_{UUT,b} \in L)$	Probability that the UUT attribute is in-tolerance
$P(\delta \in L)$	Probability that a measurement of the UUT attribute is observed to be in-tolerance
$P(e_{UUT,b} \in L, \delta \in L)$	Joint probability that a UUT attribute is in-tolerance and observed to be in-tolerance

E.2 Probability Relations
E.2.1 False Accept Risk

From the relations developed in Chapters 3 and 4, we have[64]

$$UFAR = P(\delta \in L) - P(e_{UUT,b} \in L, \delta \in L) \tag{E-1}$$

and

$$CFAR = 1 - \frac{P(e_{UUT,b} \in L, \delta \in L)}{P(\delta \in L)}. \tag{E-2}$$

E.2.2 False Reject Risk

$$FRR = P(e_{UUT,b} \in L) - P(e_{UUT,b} \in L, \delta \in L). \tag{E-3}$$

---

[63] See Appendix A, reference [1] or reference [2].

[64] To compute true in-tolerance probability from a reported in-tolerance probability, the acceptance limits A_1 and A_2 must be set equal to the tolerance limits L_1 and L_2.

E.3   True vs. Reported

Let R_{obs} denote the observed ("reported") UUT in-tolerance probability. Then R_{obs} is just $P(\delta \in \mathsf{L})$:

$$R_{obs} = P(\delta \in \mathsf{L}). \qquad (E\text{-}4)$$

Likewise, the true UUT attribute in-tolerance probability R_{true} is just

$$R_{true} = P(\delta \in \mathsf{L}). \qquad (E\text{-}5)$$

Then the risk relations can be written

$$UFAR = R_{obs} - P(e_{UUT,b} \in \mathsf{L}, \delta \in \mathsf{L}), \qquad (E\text{-}6)$$

$$CFAR = 1 - \frac{P(e_{UUT,b} \in \mathsf{L}, \delta \in \mathsf{L})}{R_{obs}}, \qquad (E\text{-}7)$$

and

$$FRR = R_{true} - P(e_{UUT,b} \in \mathsf{L}, \delta \in \mathsf{L}). \qquad (E\text{-}8)$$

From the last expression, we have

$$P(e_{UUT,b} \in \mathsf{L}, \delta \in \mathsf{L}) = R_{true} - FRR, \qquad (E\text{-}9)$$

which yields

$$R_{true} = R_{obs} - UFAR + FRR, \qquad (E\text{-}10)$$

and

$$R_{true} = R_{obs}(1 - CFAR) + FRR. \qquad (E\text{-}11)$$

E.4   Computing R_{obs} and R_{true}

The following computations involve setting tolerance limits around the UUT attribute's "true" value. Measured values δ are assumed to be normally distributed with mean $e_{UUT,b}$ and standard deviation u_{cal}. The conditional pdf for measured value δ given a UUT bias $e_{UUT,b}$ is written

$$f(\delta \mid e_{UUT,b}) = \frac{1}{\sqrt{2\pi}u_{cal}} e^{-(\delta - e_{UUT,b})^2 / 2u_{cal}^2} \qquad (E\text{-}12)$$

for each distribution discussed below. The pdf for observed in-tolerance values is denoted $f(\delta)$ in all cases. The pdf $f(\delta)$ is obtained using the expression

$$
\begin{aligned}
f(\delta) &= \int_{-\infty}^{\infty} f(e_{UUT,b}, \delta)\,de_{UUT,b} \\
&= \int_{-\infty}^{\infty} f(\delta \mid e_{UUT,b}) f(e_{UUT,b})\,de_{UUT,b},
\end{aligned}
\qquad (E\text{-}13)
$$

where $f(e_{UUT,b})$ represents the pdf for the UUT attribute bias.

The UUT attribute pdfs discussed include those for the normal, uniform, triangular, quadratic, cosine, U and lognormal distributions.

Selecting the UUT Attribute Distribution

The uniform and triangular distributions are included in this appendix only because they are mentioned in the GUM [E-2]. From a physical perspective, they are not applicable to attribute biases. It is strongly urged that other distributions be applied. Some general guidelines for selecting distributions are given in Table E-2.

Table E-2. Selection Rules for UUT Attribute Bias Distributions.

Distribution	Condition
Normal	Unless information to the contrary is available, the normal distribution should be applied as the default distribution.
Cosine	A good distribution to use if a set of containment limits is available, and values of the UUT bias exhibits a central tendency.
Lognormal	If it is suspected that the distribution of the value of interest is skewed, apply the lognormal distribution.
U (U-Shaped)	An appropriate distribution for attribute biases that vary sinusoidally with time between physical limits that are symmetric around the distribution mean or mode value.
Quadratic	A good distribution to use if a set of containment limits is available, and values of the UUT bias are widely spread around central value.
Triangular	The triangular distribution may be applicable to estimating uncertainty due to interpolation errors and, under certain circumstances, when dealing with attribute biases following testing or calibration.
Uniform (Rectangular)	An applicable distribution for the resolution error of a digital readout and for estimating the uncertainty due to quantization error or the uncertainty in RF phase angle.

E.4.1 Normally Distributed UUT Attribute Biases

Let the UUT attribute bias x be normally distributed with mean zero and standard deviation $u_{UUT,b}$, and let $-L_1$ and L_2 be the lower and upper tolerance limits for x, respectively. Then the pdf for observed UUT attribute values is

$$f(\delta) = \frac{1}{\sqrt{2\pi}u_{obs}} e^{-\delta^2/2u_{obs}^2} .$$

(E-14)

where

$$u_{obs} = \sqrt{u_{UUT,b}^2 + u_{cal}^2} ,$$

(E-15)

and where, u_{cal} is estimated in accordance with the guidelines given in Chapter 3.

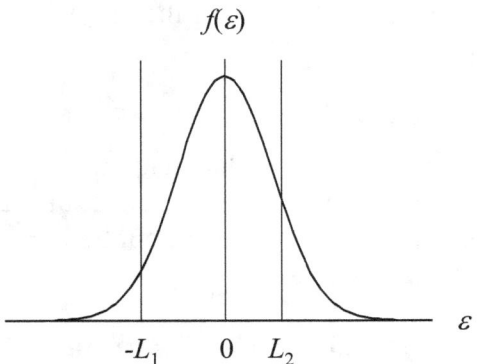

Figure E-1. The Normal Distribution.

Shown is a case where the bias ε is normally distributed and $L_1 \neq L_2$.

Using this pdf yields the observed in-tolerance probability

$$R_{obs} = \Phi\left(\frac{L_1}{u_{obs}}\right) + \Phi\left(\frac{L_2}{u_{obs}}\right) - 1 .$$ (E-16)

For a normally distributed x, if R_{true} is known, the quantity $u_{UUT,b}$ is obtained from

$$R_{true} = \Phi\left(\frac{L_1}{u_{UUT,b}}\right) + \Phi\left(\frac{L_2}{u_{UUT,b}}\right) - 1 .$$ (E-17)

E.4.1.1 Estimating R_{obs} from R_{true}

E.4.1.1.1 Case 1: $L_1 \neq L_2$

The first step is to solve for $u_{UUT,b}$ in Eq. (E-17) numerically using an iterative algorithm such as the bisection algorithm given in Appendix F. Next, compute u_{obs} using Eq. (E-15). Finally, compute R_{obs} using Eq. (E-16).

E.4.1.1.2 Case 2: $L_1 = L_2$

If $L_1 = L_2 \equiv L$, then $u_{UUT,b}$ is computed from

$$u_{UUT,b} = \frac{L}{\Phi^{-1}\left(\dfrac{1 + R_{true}}{2}\right)} ,$$ (E-18)

u_{obs} is computed using Eq. (E-15), and R_{obs} is computed using Eq. (E-16).

E.4.1.2 Estimating R_{true} from R_{obs}

E.4.1.2.1 Case 1: $L_1 \neq L_2$

The first step is to solve for u_{obs} in Eq. (E-16) numerically using an iterative algorithm such as the bisection algorithm given in Appendix F. Next, compute $u_{UUT,b}$ using Eq. (E-15). Finally, compute R_{true} using Eq. (E-17).

E.4.1.2.2 Case 2: $L_1 = L_2$

If $L_1 = L_2 \equiv L$, then u_{obs} is computed from

$$u_{obs} = \frac{L}{\Phi^{-1}\left(\dfrac{1 + R_{obs}}{2}\right)},$$

$u_{UUT,b}$ is then computed using Eq. (E-15), and R_{true} is computed using Eq. (E-17).

E.4.2 Uniformly Distributed Attribute Biases
E.4.2.1 Estimating R_{obs} from R_{true}

In cases where UUT attribute biases are uniformly distributed, the pdf for $e_{UUT,b}$ is written

$$f(e_{UUT,b}) = \begin{cases} \dfrac{1}{2a}, & -a \le x \le a \\ 0, & \text{otherwise,} \end{cases} \tag{E-19}$$

where $\pm a$ are the bounding limits for the distribution as shown in Figure E-2.

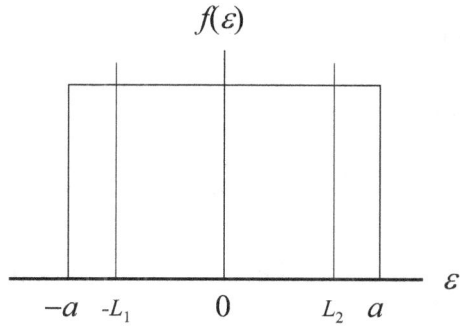

Figure E-2. The Uniform Distribution

The tolerance limits $\pm L$ and the quantities R_{true} or R_{obs} are used to solve for the limiting values $\pm a$.

Applying Eqs. (E-12) and (E-13) with Eq. (19) yields the pdf $f(\delta)$

$$\begin{aligned} f(\delta) &= \int_{-\infty}^{\infty} f(\delta \mid e_{UUT,b}) f(e_{UUT,b}) de_{UUT,b} \\ &= \frac{1}{2a\sqrt{2\pi}u_{cal}} \int_{-a}^{a} e^{-(\delta - e_{UUT,b})^2/2u_{cal}^2} de_{UUT,b} \\ &= \frac{1}{2a\sqrt{2\pi}} \int_{(\delta-a)/u_{cal}}^{(\delta+a)/u_{cal}} e^{-\zeta^2/2} d\zeta \\ &= \frac{1}{2a}\left[\Phi\left(\frac{\delta+a}{u_{cal}}\right) - \Phi\left(\frac{\delta-a}{u_{cal}}\right)\right]. \end{aligned} \tag{E-20}$$

The quantity R_{obs} is obtained by integrating Eq. (E-20) from $-L_1$ to L_2

$$R_{obs} = \frac{1}{2a} \int_{-L_1}^{L_2} \left[\Phi\left(\frac{\delta + a}{u_{cal}}\right) - \Phi\left(\frac{\delta - a}{u_{cal}}\right) \right] d\delta .$$ (E-21)

Applications of the uniform distribution typically involve attributes with symmetric tolerance limits, where $L_1 = L_2 = L$. Then, for uniformly distributed UUT attribute biases, we have

$$R_{true} = \frac{L}{a} .$$ (E-22)

Substituting $a = L / R_{true}$ in Eq. (E-21) then gives

$$R_{obs} = \frac{1}{2a} \int_{-L}^{L} \left[\Phi\left(\frac{\delta + L / R_{true}}{u_{cal}}\right) - \Phi\left(\frac{\delta - L / R_{true}}{u_{cal}}\right) \right] d\delta .$$ (E-23)

This expression is computed using a numerical integration routine. See, for example, the routine provided in Appendix F.

E.4.2.2 Estimating R_{true} from R_{obs}

Estimating R_{true} from R_{obs} makes use of Eq. (E-23). This expression is used to solve for R_{true}, given R_{obs}, using the bisection method of Appendix F with the integration computed numerically at each step. Note that, since $R_{true} \geq R_{obs}$, the bracketing values are R_{obs} and 1.0. The initial value could be something like $(1 + R_{obs}) / 2$.

E.4.3 Triangularly Distributed Attribute Biases

E.4.3.1 Estimating R_{obs} from R_{true}

In cases where UUT attribute biases follow the triangular distribution, the pdf for $e_{UUT|b}$ is given by

$$f(e_{UUT,b}) = \begin{cases} (a + e_{UUT,b}) / a^2, & -a \leq e_{UUT,b} \leq 0 \\ (a - e_{UUT,b}) / a^2, & 0 \leq e_{UUT,b} \leq a \\ 0, & \text{otherwise.} \end{cases}$$ (E-24)

where $\pm a$ are the bounding limits for the distribution as shown in Figure E.3.

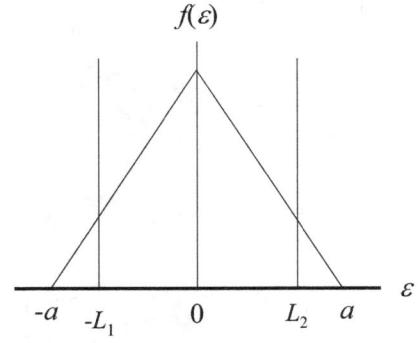

Figure E-3. The Triangular Distribution

The quantities L and R_{true} or R_{obs} are used to solve for the limiting values $\pm a$.

Like the uniform distribution, applications of the triangular distribution typically involve attributes with symmetric tolerance limits, where $L_1 = L_2 = L$. Then, applying Eqs. (E-12) and (E-13) with (E-24) yields the pdf $f(\delta)$

$$
\begin{aligned}
f(\delta) &= \int_{-\infty}^{\infty} f(\delta \mid \zeta) f(\zeta) d\zeta \\
&= \frac{1}{\sqrt{2\pi} a^2 u_{cal}} \left[\int_{-a}^{0} (a+\zeta) e^{-(\delta-\zeta)^2/2u_{cal}^2} d\zeta + \int_{0}^{a} (a-\zeta) e^{-(\delta-\zeta)^2/2u_{cal}^2} d\zeta \right].
\end{aligned}
\tag{E-25}
$$

where ζ is a dummy integration variable for $e_{UUT,b}$. The quantity R_{obs} is obtained by integrating Eq. (E-25) from $-L$ to L.

$$
\begin{aligned}
R_{obs} &= \frac{1}{\sqrt{2\pi} a^2 u_{cal}} \left[\int_{-a}^{0} d\zeta (a+\zeta) \int_{-L}^{L} d\delta\, e^{-(\delta-\zeta)^2/2u_{cal}^2} + \int_{0}^{a} d\zeta (a-\zeta) \int_{-L}^{L} d\delta\, e^{-(\delta-\zeta)^2/2u_{cal}^2} \right] \\
&= \frac{2}{a^2} \int_{0}^{a} \left[\Phi\left(\frac{L+\zeta}{u_{cal}}\right) + \Phi\left(\frac{L-\zeta}{u_{cal}}\right) \right] (a-\zeta) d\zeta - 1.
\end{aligned}
\tag{E-26}
$$

For triangularly distributed UUT attribute biases, we have

$$
a = \frac{L}{R_{true}} \left(1 + \sqrt{1 - R_{true}}\right).
\tag{E-27}
$$

Substituting this relation for a in Eq. (E-26) then gives R_{obs}. The expression for R_{obs} is computed numerically using an integration routine, such as is described in Appendix F.

E.4.3.2 Estimating R_{true} from R_{obs}

Estimating R_{true} from R_{obs} makes use of Eqs. (E-26) and (E-27). The resulting expression is used to solve for R_{true} using the bisection method of Appendix F with the integration computed numerically at each step. Note that, since $R_{true} \geq R_{obs}$, the bracketing values are R_{obs} and 1.0. The initial value could be something like $(1 + R_{obs}) / 2$.

E.4.4 Quadratically Distributed Attribute Biases

E.4.4.1 Estimating R_{obs} from R_{true}

In cases where UUT attribute biases follow the quadratic distribution, the pdf for $e_{UUT,b}$ is given by

$$
f(x) = \begin{cases} \dfrac{3}{4a}\left[1-(e_{UUT,b}/a)^2\right], & -a \leq x \leq a \\ 0, & \text{otherwise}, \end{cases}
\tag{E-28}
$$

where $\pm a$ are the bounding limits for the distribution as shown in Figure E.4.

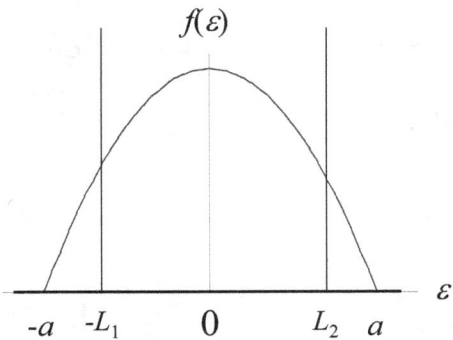

Figure E-4. The Quadratic Distribution

The quantities L and R_{true} or R_{obs} are used to solve for the limiting values $\pm a$.

Applications of the quadratic distribution typically involve attributes with symmetric tolerance limits, where $L_1 = L_2 = L$. Then, applying Eqs. (E-12) and (E-13) with (E-28) yields the pdf $f(\delta)$

$$f(\delta) = \int_{-\infty}^{\infty} f(\delta \mid e_{UUT,b}) f(e_{UUT,b}) de_{UUT,b}$$

$$= \frac{3}{4a\sqrt{2\pi}u_{cal}} \int_{-a}^{a} \left[1 - (e_{UUT,b}/a)^2\right] e^{-(\delta - e_{UUT,b})^2/2u_{cal}^2} de_{UUT,b}.$$

(E-29)

The quantity R_{obs} is obtained by integrating Eq. (E-29) from $-L$ to L.

$$R_{obs} = \frac{3}{4a\sqrt{2\pi}u_{cal}} \int_{-a}^{a} \left[1 - (e_{UUT,b}/a)^2\right] de_{UUT,b} \int_{-L}^{L} e^{-(\delta - e_{UUT,b})^2/2u_{cal}^2} dy$$

$$= \frac{3}{4a} \int_{-a}^{a} \left[\Phi\left(\frac{L + e_{UUT,b}}{u_{cal}}\right) + \Phi\left(\frac{L - e_{UUT,b}}{u_{cal}}\right) \right] \left[1 - (e_{UUT,b}/a)^2\right] de_{UUT,b} - 1.$$

(E-30)

For quadratically distributed UUT attribute biases with symmetric tolerance limits, we have

$$a = \frac{L}{2R_{true}} \left(1 + 2\cos\left[\frac{1}{3}\arccos(1 - 2R_{true}^2)\right]\right).$$

(E-31)

Substituting this relation for a in Eq. (E-30) then gives R_{obs}. The expression for R_{obs} is computed numerically using an integration routine, such as is described in Appendix F.

E.4.4.2 Estimating R_{true} from R_{obs}

Estimating R_{true} from R_{obs} makes use of Eqs. (E-26) and (E-27). The resulting expression is used to solve for R_{true} using the bisection method of Appendix F with the integration computed numerically at each step. Note that, since $R_{true} \geq R_{obs}$, the bracketing values are R_{obs} and 1.0. The initial value could be something like $(1 + R_{obs}) / 2$.

E.4.5 Cosine Distributed Attribute Biases
E.4.5.1 Estimating R_{obs} from R_{true}

In cases where UUT attribute biases follow the cosine distribution, the pdf for $e_{UUT,b}$ is given by

$$f(e_{UUT,b}) = \begin{cases} \dfrac{1}{2a}\left[1+\cos(\pi e_{UUT,b}/a)\right], & -a \leq e_{UUT,b} \leq a \\ 0, & \text{otherwise}, \end{cases}$$

(E-32)

where $\pm a$ are the bounding limits for the distribution as shown in Figure E.5.

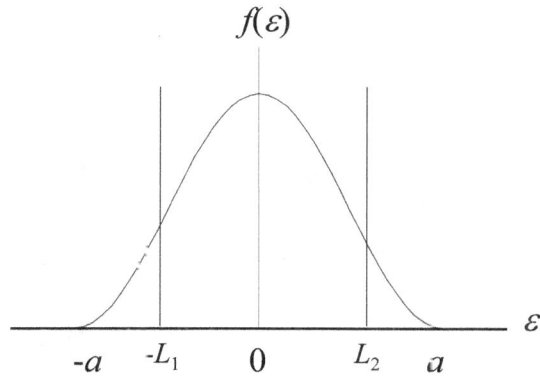

$f(\varepsilon)$

$-a \quad -L_1 \quad 0 \quad L_2 \quad a \qquad \varepsilon$

Figure E-5. The Cosine Distribution

The quantities L and R_{true} or R_{obs} are used to solve for the limiting values $\pm a$.

Applications of the cosine distribution typically involve attributes with symmetric tolerance limits, where $L_1 = L_2 = L$. Then, applying Eqs. (E-12) and (E-13) with (E-32) yields the pdf $f(\delta)$

$$f(\delta) = \int_{-\infty}^{\infty} f(\delta \mid e_{UUT,b})f(e_{UUT,b})de_{UUT,b}$$

$$= \frac{1}{2a\sqrt{2\pi}u_{cal}} \int_{-a}^{a}\left[1+\cos(\pi e_{UUT,b}/a)\right]e^{-(\delta-e_{UUT,b})^2/2u_{cal}^2}de_{UUT,b}.$$

(E-33)

The observed in-tolerance probability R_{obs} is obtained by integrating Eq. (E-33) from $-L$ to L.

$$R_{obs} = \frac{1}{2a\sqrt{2\pi}u_{cal}} \int_{-a}^{a} de_{UUT,b}\left[1+\cos(\pi e_{UUT,b}/a)\right]\int_{-L}^{L} d\delta\, e^{-(\delta-e_{UUT,b})^2/2u_{cal}^2}$$

$$= \frac{1}{2a} \int_{-a}^{a}\left[\Phi\left(\frac{L+e_{UUT,b}}{u_{cal}}\right)+\Phi\left(\frac{L-e_{UUT,b}}{u_{cal}}\right)\right]\left[1+\cos(\pi e_{UUT,b}/a)\right]de_{UUT,b}-1.$$

(E-34)

For quadratically distributed UUT attribute biases with symmetric tolerance limits, R_{true} is given by

$$R_{true} = \frac{L}{a}+\frac{1}{\pi}\sin(\pi L/a)$$

(E-35)

- 153 -

The distribution limit a is solved for by numerical iteration. Substituting the solution for a in Eq. (E-34) then gives R_{obs}. The expression for R_{obs} is computed by numerical integration. Both the iteration and integration routines in Appendix F have been found to be effective in obtaining solutions.

E.4.5.2   Estimating R_{true} from R_{obs}

Estimating R_{true} from R_{obs} makes use of Eqs. (E-34) and (E-35). The resulting expression is used to solve for R_{true} using the bisection method of Appendix F with the integration computed numerically at each step. Note that, since $R_{true} \geq R_{obs}$, the bracketing values are R_{obs} and 1.0. The initial value could be something like $(1 + R_{obs}) / 2$.

E.4.6   U-Distributed Attribute Biases
E.4.6.1   Estimating R_{obs} from R_{true}

In cases where UUT attribute biases follow the U distribution, the pdf for $e_{UUT|b}$ is given by

$$f(e_{UUT,b}) = \begin{cases} \dfrac{1}{\pi\sqrt{a^2 - e_{UUT,b}^2}}, & -a \leq e_{UUT,b} \leq a \\ 0, & \text{otherwise}, \end{cases} \tag{E-36}$$

where $\pm a$ are the bounding limits for the distribution as shown in Figure E.6.

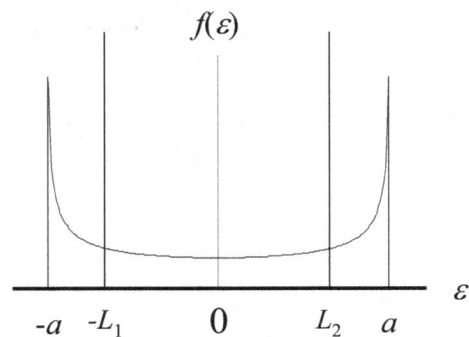

Figure E-6. The U Distribution

The quantities L and R_{true} or R_{obs} are used to solve for the limiting values $\pm a$.

Applications of the U distribution typically involve attributes with symmetric tolerance limits, where $L_1 = L_2 = L$. Then, applying Eqs. (E-12) and (E-13) with (E-36) yields the pdf $f(y)$

$$\begin{aligned} f(\delta) &= \int_{-\infty}^{\infty} f(\delta \mid e_{UUT,b}) f(\delta) d\delta \\ &= \frac{1}{\pi\sqrt{2\pi}u_{cal}} \int_{-a}^{a} e^{-(\delta - e_{UUT,b})^2/2u_{cal}^2} \frac{de_{UUT,b}}{\sqrt{a^2 - e_{UUT,b}^2}}. \end{aligned} \tag{E-37}$$

The observed in-tolerance probability R_{obs} is obtained by integrating Eq. (E-37) from $-L$ to L.

$$R_{obs} = \frac{1}{\pi\sqrt{2\pi}u_{cal}} \int_{-a}^{a} \frac{de_{UUT,b}}{\sqrt{a^2 - e_{UUT,b}^2}} \int_{-L}^{L} d\delta e^{-(\delta - e_{UUT,b})^2/2u_{cal}^2}$$

$$= \frac{1}{\pi} \int_{-a}^{a} \left[\Phi\left(\frac{L + e_{UUT,b}}{u_{cal}} \right) + \Phi\left(\frac{L - e_{UUT,b}}{u_{cal}} \right) \right] \frac{de_{UUT,b}}{\sqrt{a^2 - e_{UUT,b}^2}} - 1.$$

(E-38)

For U-distributed UUT attribute biases with symmetric tolerance limits, a is given by

$$a = \frac{L}{\sin\left(\pi R_{true} / 2\right)}.$$

(E-39)

The distribution limit a is solved for by numerical iteration. Substituting the solution for a in Eq. (E-38) then gives R_{obs}. The expression for R_{obs} is computed by numerical integration. The integration routine in Appendix F has been found to be effective in obtaining solutions.

E.4.6.2 Estimating R_{true} from R_{obs}

Estimating R_{true} from R_{obs} makes use of Eqs. (E-38) and (E-39). The resulting expression is used to solve for R_{true} using the bisection method of Appendix F with the integration computed numerically at each step. Note that, since $R_{true} \geq R_{obs}$, the bracketing values are R_{obs} and 1.0. The initial value could be something like $(1 + R_{obs}) / 2$.

E.4.7 Lognormally Distributed Attribute Biases
E.4.7.1 The Distribution

Achieving solutions for R_{obs} and R_{true} for lognormally distributed attribute biases is made easier if we work with distributions of attribute values rather than biases in these values. Hence we work with the distributions shown in Figure E-7 and E-8.

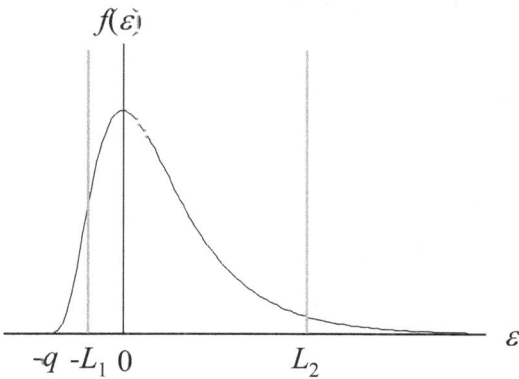

Figure E-7. The Right-Handed Lognormal Distribution

The variable $e_{UUT,b}$ represents UUT attribute biases. The parameters $-L_1$ and L_2 are attribute tolerance limits. The mode value for $e_{UUT,b}$ is 0 and the limiting value for the distribution is $-q$. L_1, L_2 and q characterize the distribution. The attribute q is ordinarily computed using R_{true}. Solutions for the parameters of the lognormal distribution are given in [E-3] and [E-4].

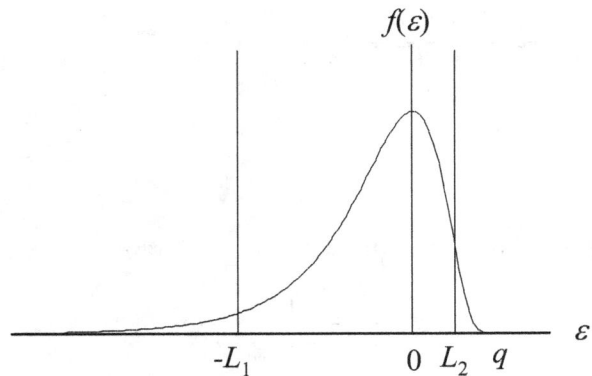

Figure E-8. The Left-Handed Lognormal Distribution

With the left-handed lognormal distribution, the distribution limit > 0.

The following treatment focuses on UUT attribute values that follow a right-handed lognormal distribution. Using the results to accommodate left-handed distributions involves merely applying transformations that are discussed later.

The right-handed lognormal pdf for $e_{UUT,b}$ is given by

$$f(e_{UUT,b}) = \frac{1}{\sqrt{2\pi}\lambda(e_{UUT,b}+q)} \exp\left\{-\left[\ln\left(\frac{e_{UUT,b}+q}{m+q}\right)\right]^2 / 2\lambda^2\right\} \tag{E-40}$$

where m is the median value of the distribution and λ is a "shape" attribute. The pdf for the left-handed lognormal is the mirror image of the pdf for the right-handed distribution.

Applying Eqs. (E-12) and (E-13) with (E40) yields the pdf $f(\delta)$

$$f(\delta) = \int_{-\infty}^{\infty} f(\delta \mid e_{UUT,b}) f(e_{UUT,b}) de_{UUT,b}$$

$$= \frac{1}{2\pi\lambda u_{cal}} \int_{q}^{\infty} e^{-(\delta-e_{UUT,b})^2/2u_{cal}^2} \exp\left\{-\left[\ln\left(\frac{e_{UUT,b}+q}{m+q}\right)\right]^2 / 2\lambda^2\right\} \frac{de_{UUT,b}}{(e_{UUT,b}+q)} \tag{E-41}$$

$$= \frac{1}{2\pi u_{cal}} \int_{-\infty}^{\infty} e^{-(\delta-\alpha)^2/2u_{cal}^2} e^{-\zeta^2/2} d\zeta,$$

where

$$\alpha = (m+q)e^{\lambda\zeta} - q. \tag{E-42}$$

E.4.7.2   Observed In-Tolerance Probability

The observed in-tolerance probability R_{obs} is obtained by integrating $f(y)$ from $-L_1$ to L_2.

$$R_{obs} = \frac{1}{2\pi u_{cal}} \int_{-\infty}^{\infty} e^{-\zeta^2/2} d\zeta \int_{-L1}^{L2} e^{-(\delta-\alpha)^2/2u_{cal}^2} d\delta$$

$$= \frac{1}{2\pi} \int_{-\infty}^{\infty} d\zeta e^{-\zeta^2/2} \int_{-(L_1+\alpha)/u_{cal}}^{(L_2-\alpha)/u_{cal}} d\xi e^{-\xi^2/2}$$

$$= \frac{1}{\sqrt{2\pi}} \int_{-\infty}^{\infty} \left[\Phi\left(\frac{L_2-\alpha}{u_{cal}}\right) + \Phi\left(\frac{L_1+\alpha}{u_{cal}}\right) - 1 \right] e^{-\zeta^2/2} d\zeta$$

$$= \frac{1}{\sqrt{2\pi}} \int_{-\infty}^{\infty} \left[\Phi\left(\frac{L_2-\alpha}{u_{cal}}\right) + \Phi\left(\frac{L_1+\alpha}{u_{cal}}\right) \right] e^{-\zeta^2/2} d\zeta - 1,$$

(E-43)

where α is given in Eq. (E-42).

E.4.7.3 True In-Tolerance Probability

The integral for R_{obs} is obtained by numerical iteration. The integration routine in Appendix F has been found to be effective in obtaining solutions. Solutions for q, m and λ are computed from R_{true}, computed by integrating $f(e_{UUT,b})$ in Eq. (E-40) from $-L_1$ to L_2

$$R_{true} = \frac{1}{\sqrt{2\pi}\lambda} \int_{-L_1}^{L_2} \exp\left\{ -\left[\ln\left(\frac{e_{UUT,b}+q}{m+q}\right)\right]^2 / 2\lambda^2 \right\} \frac{de_{UUT,b}}{(e_{UUT,b}+q)}$$

$$= \frac{1}{\sqrt{2\pi}} \int_{\ln[-(L_1-q)/(m+q)]/\lambda}^{\ln[(L_2+q)/(m+q)]/\lambda} e^{-\zeta^2/2} d\zeta$$

(E-44)

$$= \Phi\left(\frac{\ln[(L_2+q)/(m+q)]}{\lambda}\right) - \Phi\left(\frac{\ln[-(L_1+q)/(m+q)]}{\lambda}\right).$$

E.4.7.4 Obtaining Distribution Parameters

In solving for q, m and λ, it is assumed that L_1 and L_2 are known, along with the true in-tolerance probability R_{true}. The details are given in [E-3] and [E-4]. The relationships between the parameters of the distribution are shown in Table E-3.

Table E-3. Parameters of the Lognormal Distribution.

Parameter	Characteristics
Mode	0
Mean ($\bar{e}_{UUT,b}$)	$(m+q)e^{\lambda^2/2} - q$
Median (m)	$(\mu+q)e^{\lambda^2} - q$
Variance (u_{cal}^2)	$(m+q)^2 e^{\lambda^2}(e^{\lambda^2}-1)$
Standard Deviation (u_{cal})	$|m+q|e^{\lambda^2/2}\sqrt{e^{\lambda^2}-1}$

From the foregoing, we see that, if q is known, λ can be solved for numerically using the bisection algorithm of Appendix F along with the relation for the median shown in Table E-3. Likewise, if λ is known, q can be solved for in the same way.

In many cases, none of the parameters q, λ, or m is known. In these cases, an attempt can be made to solve for them, provided we know L_1, L_2 and R_{true}. The details are given in [E-3] and [E-4].

E.4.7.5 Estimating R_{true} from R_{obs}

If any of two parameters of Eq. (E-42) are known, Eq. (E-43) can be used to iteratively solve for the third attribute. For instance, if q and m are known, λ can be obtained using the bisection algorithm of Appendix F. Once all three parameters are known, R_{true} can be readily computed from Eq. (E-44).

E.5 Examples

Tables E-4 and E-5 provide comparisons of true vs. reported in-tolerance probabilities. The table entries were generated using the algorithms of Appendix F.

Table E-4. True vs. Reported % In-Tolerance for a Reported 95% In-Tolerance.

Distribution	Lower Tol. Limit	Upper Tol. Limit	Cal Uncertainty	True % In-Tolerance
Normal	-10	10	1.2755	95.7054
Uniform	-10	10	1.2755	No Solution
Triangular	-10	10	1.2755	96.0366
Quadratic	-10	10	1.2755	96.4087
Cosine	-10	10	1.2755	95.9962
U	-10	10	1.2755	No Solution
Lognormal	-10	20	1.2755	95.4813

Table E-5. Reported vs. True % In-Tolerance for a True 95% In-Tolerance.

Distribution	Lower Tol. Limit	Upper Tol. Limit	Cal Uncertainty	Reported % In-Tolerance
Normal	-10	10	1.2755	94.2757
Uniform	-10	10	1.2755	92.2601
Triangular	-10	10	1.2755	94.0219
Quadratic	-10	10	1.2755	93.7123
Cosine	-10	10	1.2755	94.0183
U	-10	10	1.2755	88.8954
Lognormal	-10	20	1.2755	94.5074

Appendix E References

[E-1] Ferling, J., "The Role of Accuracy Ratios in Test and Measurement Processes," *Proc. Meas. Sci. Conf.*, Long Beach, January 1984.

[E-2] ANSI/NCSL Z540-2-1997, *U.S. Guide to the Expression of Uncertainty in Measurement*, National Conference of Standards Laboratories, Boulder, 1997.

[E-3] NCSLI, Determining *and Reporting Measurement Uncertainties*, Recommended Practice RP-12, National Conference of Standards Laboratories, April 1995.

[E-4] NASA, *Measurement Uncertainty Analysis Principles and Methods*, KSC-UG-2809, National Aeronautics and Space Administration, November 2007.

Appendix F: Useful Numerical Algorithms

This appendix provides routines that have been found useful in building measurement decision risk analysis programs and other programs relating to analytical measurement science. They are presented in the Visual Basic 6 programming language.

F.1 Bisection Routine

F.1.1 Function Root

This routine solves for the root of a function computed for a function called from a routine referred to as "Fun." The variables x1 and x2 are bracketing quantities established with a subroutine named "Bracket." The vector p() contains the parameters of the called function.

```
    Dim iMax As Integer, i As Integer
    Dim dx As Double, f As Double, fMid As Double, xMid As
Double
    Dim eps As Double, x1 As Double, x2 As Double, rt As Double

'   Set the precision of the estimate
    eps = 0.0000000001

'   Get bracketing values x1 and x2
'   Initial values
    If x < 0 Then
        x1 = 2 * x
        x2 = x / 2
    Else
        x1 = x / 2
        x2 = 2 * x
    End If
    Do While True
        Bracket x1, x2, p(), Fail
        If Fail Then
            If x1 < 0 Then
                x1 = 2 + x1
                x2 = x2 / 2
            Else
                x1 = x1 / 2
                x2 = 2 * x2
            End If
        Else
            Exit Do
        End If
    Loop

    If Not Fail Then
        iMax = 100
        fMid = Fun(x2, p())
```

```
            f = Fun(x1, p())
            If f < 0 Then
                    rt = x1
                    dx = x2 - x1
            Else
                    rt = x2
                    dx = x1 - x2
            End If
            For i = 1 To iMax
                    dx = dx / 2
                    xMid = rt + dx
                    fMid = Fun(xMid, p())
                    If fMid <= 0 Then rt = xMid
                    If (Abs(dx) < eps Or fMid = 0) Then Exit For
            Next
            Root = rt
    End If
```

F.1.2   Function Bracket

This routine finds bracketing values for a function computed in Function "Fun." The brackets
are used in the function "Root." The routine is adapted from [F-1]. The variables x1, x2 and the
parameter vector for the function of interest are passed into this function.

```
    Dim nTry As Integer, i As Integer
    Dim factor As Double, f1 As Double, f2 As Double

'   Set the initial parameters
    factor = 1.6
    nTry = 50
    f1 = Fun(x1, p())
    f2 = Fun(x2, p())

    Fail = True
    For i = 1 To nTry
        If f1 * f2 < 0 Then 'have bracketing values
            Fail = False
            Exit For
        End If
        If Abs(f1) < Abs(f2) Then
            x1 = x1 + factor * (x1 - x2)
            f1 = Fun(x1, p())
        Else
            x2 = x2 + factor * (x2 - x1)
            f2 = Fun(x2, p())
        End If
    Next
```

F.2   Gauss Quadrature Integration[65]

The following routine has been useful when integrating statistical and other mathematical functions. The routine is written in VB6 language for simplicity. The routine integrates a function (fun) between the limits L1 and L2 using Gauss quadrature. The computations are performed at n points using abscissa and weight values obtained from the GaussLegendre routine described in Section F.2.2.

F.2.1   Subroutine GaussQuadrature

```
Sub GaussQuadrature(L1 As Double, L2 As Double, p()as Double,
fInt As Double)

'  Returns fInt as the integral obtained by n-point Gauss-
Legendre
'  integration of the function fun between the integration
limits
'  L1 and L2.  The function fun is evaluated n times
'  at interior points in the range of integration.

    'p()      - Parameters of the function
    'fInt     - The integral of the function.

    Dim i As Integer, n As Integer

    n = 40 'the number of weights and abscissas used in the
integration

'  Calculate the abscissas and weights each time a function
'  is integrated.
'  Abscissas and weights are a function of the variables n, L2
and L1.

    ReDim Abscissa(n)
    ReDim Weights(n)
    GaussLegendre L1, L2, Abscissa(), Weights(), (n)      'Get the
abscissa
    and weight arrays.

'  Perform the "integration:
    fInt = 0
    For i = 1 To n
        fInt = fInt + Weights(i) * fun(p()) * Abscissa(i)
    Next

    Exit Sub
```

---

[65] Adapted from [F-1].

F.2.2 Subroutine GaussLegendre

```
Sub GaussLegendre(L1 As Double, L2 As Double, Abscissa() As
Double, Weights() As Double, n As Integer)

'  Routine returns abscissa and weight arrays for an n-point
Gauss-Legendre
   integration of a
'  function used in subroutine GaussianQuadrature.  This routine
needs
   high precision.

   Dim i As Integer, j As Integer
   Dim eps As Double, xM As Double, xL As Double, z As Double,
z1 As Double
   Dim p1 As Double, p2 As Double, p3 As Double, pp As Double

   eps = 0.00000000000003 'The precision of the estimates
   xM = (L1 + L2) / 2
   xL = (L2 - L1) / 2

   ReDim Abscissa(n)
   ReDim Weights(n)

   For i = 1 To n
      z = Cos(pi * (i - 0.25) / (n + 0.5))
'     Starting with the above approximation to the ith root,
'     we solve for the Legendre polynomial p1 using the
'     Newton-Raphson method.
      Do While True
            p1 = 1
            p2 = 0
'     Loop the recurrence relation to get the Legendre
polynomial        '      evaluated at z.
            For j = 1 To n
                  p3 = p2
                  p2 = p1
                  p1 = ((2 * j - 1) * z * p2 - (j - 1) * p3) / j
            Next
'     p1 is now the desired Legendre polynomial.  Next
compute pp,
'      its derivative, by
'      a standard relation involving p2, the polynomial of
'      one order lower than p1.
            pp = n * (z * p1 - p2) / (z * z - 1)
            z1 = z
            z = z1 - p1 / pp
```

```
              If Abs(z - z1) <= eps Then Exit Do
        Loop
   '  Scale the root to the desired interval.
        Abscissa(i) = xM - xL * z
   '  Compute the ith weight.
        Weights(i) = 2 * xL / ((1 - z * z) * pp * pp)
      Next

End Sub
```

Appendix F References

[F-1]   Press, W., Vettering, W., Teukolsky, S. and Flannery, B., *Numerical Recipes*, 2nd Ed., Cambridge University Press, Cambridge, 1992.

Appendix G: Calibration Feedback Analysis

An important problem in calibration is evaluation of the significance of an out-of-tolerance MTE attribute on workload item attributes which were previously tested with it. Common approaches to this evaluation do not quantitatively link the out-of-tolerance condition to the validity of the workload item test. The question of significance cannot be adequately answered without doing so.

This appendix presents a method for assessing the significance of an observed out-of-tolerance in a test instrument on the testing of an end item in terms of end item false accept risk. In this scenario, an attribute of a UUT end item is tested using a reference attribute of an item of MTE. The UUT attribute is found in-tolerance and is accepted. The false accept risk for the UUT attribute at the time of test is estimated from information obtained during calibration of the test MTE.

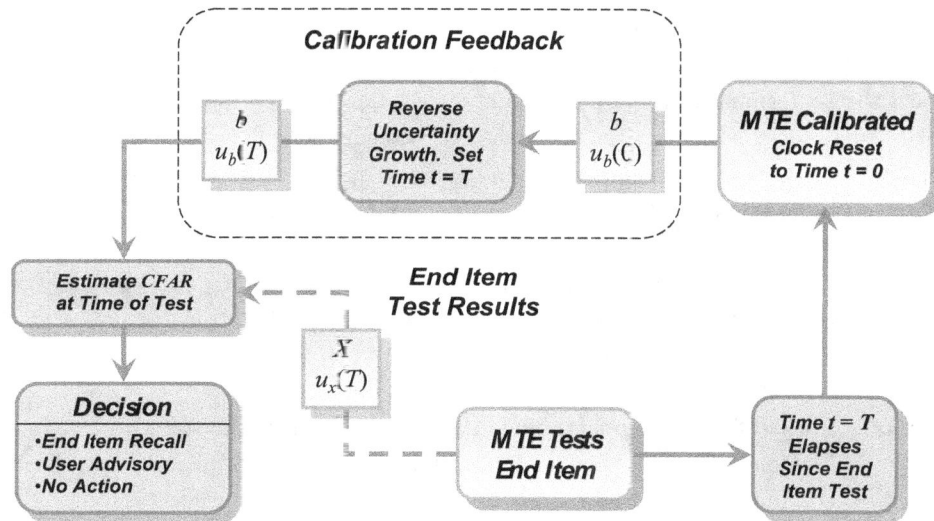

Figure G-1. The Calibration Feedback Loop.

An MTE attribute used in the test of an end item is calibrated after a time T has elapsed since the test. The results of calibration are an estimated MTE attribute bias b and an uncertainty in this estimate $u_b(0)$. An estimate of the uncertainty in b at the time of the end item test $u_b(T)$ is obtained by computing uncertainty growth over a time interval $t = T$. The quantities b, $u_b(T)$, the test result X, and the test process uncertainty $u_x(T)$ are used to compute *CFAR* for the tested end item attribute. A decision is then made whether to recall the end item, advise the user or take no action.

G.1   Appendix G Nomenclature

The nomenclature used in this appendix is defined in Table G-1.

Table G-1. Variables Used in Calibration Feedback Analysis.

Variable		Description
UUT	-	a unit under test end item
MTE	-	the measuring and test equipment used in performing the end item test
x	-	random variable representing deviations from nominal of the end item UUT attribute of interest
X	-	the value of the end item UUT attribute at the time of test
y	-	random variable representing values of x measured during testing by the MTE
Y	-	the value of X measured during testing
z	-	random variable representing the bias in the MTE at the time of its calibration
w	-	random variable representing the value of z measured during MTE calibration
Z	-	the value of z measured during calibration
u_x	-	estimated standard uncertainty in x at time of test
u_y	-	estimated standard uncertainty in y at time of test
u_z	-	estimated standard uncertainty in z at the time of the calibration of the MTE
u_w	-	estimated standard uncertainty in w at the time of the calibration of the MTE
b	-	MTE attribute bias estimated from calibration data
u_b	-	estimated uncertainty in b
T	-	time elapsed between UUT test and MTE calibration
$u_b(T)$	-	estimated uncertainty in b at the time of the UUT test
$\pm L$	-	tolerance limits for the UUT attribute
$f_z(z)$	-	pdf for z
$f_w(w|z)$	-	pdf for measurements of the UUT attribute made using the MTE
P_{xy}	-	joint probability that a UUT attribute will both be in-tolerance and observed to be in-tolerance
P_y	-	probability that UUT attribute values will be observed to be in-tolerance.
$CFAR$	-	false accept risk (see Chapter 3).

Note that, because both an MTE calibration and an end item test are involved, the notation in this appendix departs somewhat from that employed in the bulk of this annex. Also, the calibration feedback problem involves an elapsed time T between end item test and a subsequent MTE calibration. This means that uncertainty growth is a factor. Accordingly, some of the notation of Appendix J is employed in the discussion.

G.2   Estimating Risk from a Measured UUT Value

A UUT attribute is tested at time $t = 0$. The value X of the UUT is measured by the testing MTE to be Y. The MTE is later calibrated at time $t = T$ and is found to have a bias b with uncertainty $u_b(0)$. The value b and the bias uncertainty estimate $u_b(0)$ are obtained using Bayesian analysis [G-1 – G-6]. The test and calibration loop is shown in Figure G-1. The estimation process is described in Chapter 4, Appendix C and Appendix D.

We want to estimate the false accept risk associated with testing the UUT attribute with the MTE of interest at time $t = 0$. The first step in the process is to estimate the uncertainty $u_b(0)$ in b at this time. This is done by computing the uncertainty growth backward for a time interval $t = T$.

We now use Bayesian analysis to determine the false accept risk of the UUT test. We first determine the UUT bias uncertainty u_x from the UUT tolerance limits $-L$ and L and an *a priori* estimate of its in-tolerance probability at the time of test. The estimate for the quantity u_x is described in Section G.6. We next estimate the total test process uncertainty u_y, employing σ_b as the MTE bias uncertainty.

We then define a variable

$$r = \frac{u_x}{u_y},$$

and compute the false accept risk as

$$CFAR = 1 - P_{in},$$

where

$$P_{in} = \Phi(a_+) + \Phi(a_-) - 1$$

and

$$a_{\pm} = \frac{\sqrt{1+r^2}\left(L \pm \frac{(Y-b)r^2}{1+r^2}\right)}{u_x}.$$

G.3 Example

Imagine that an estimate of the bias in the 10 VDC scale of a hypothetical digital multimeter is obtained during the calibration. Suppose this attribute was used to test an attribute of a DVD player. During the test, the DVD player attribute was measured to be 4.81 mV above nominal.

The information used to develop uncertainty growth characteristics for the MTE attribute is shown in Figure G-2. The uncertainties of both the MTE calibration and the DVD test are shown in Table G-2.

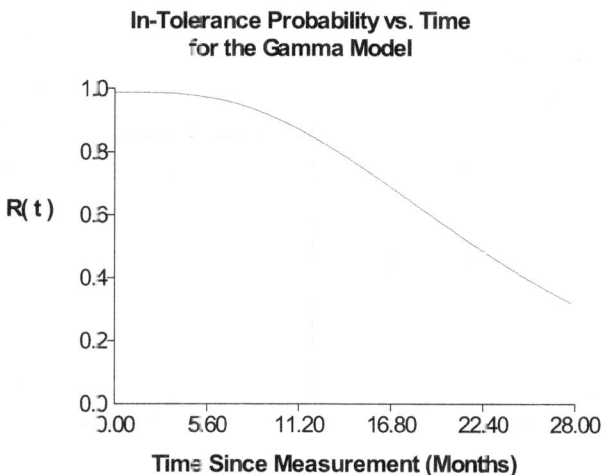

In-Tolerance Probability vs. Time
for the Gamma Model

Reliability vs. time information used to develop the uncertainty growth characteristics of the MTE attribute. The BOP and EOP in-tolerance probabilities are set at 99% and 85%, respectively. The reliability model shown is a modified gamma model with the reliability function

$$R(t) = e^{-\lambda t}[1 + \lambda t + (\lambda t)^2 / 2 + (\lambda t)^3 / 6] \quad [G-7].$$

Table G-2. Example UUT test and MTE calibration results.

End Item (UUT) Attribute Test Data

Tolerance Type:	Two-Sided	
Upper Tol Limit:	8	mV
Lower Tol Limit:	-10	mV
UUT % In-Tolerance at Time of Test:	95.00	
Estimated UUT Bias Uncertainty:	**4.4826**	mV
Deviation Recorded at Test:	**4.81**	mV
Test Process Uncertainty:	**2.33**	mV
(Excludes MTE Bias Uncertainty)		
Date of Test:	4/15/2007	
Time Elapsed Since Test:	370	Days
MTE Bias Uncertainty at Time of Test:	**0.9239**	mV

Test MTE Cal Data

Tolerance Type:	Two-Sided	
Upper Tol Limit:	2.5	mV
Lower Tol Limit:	-2.5	mV
MTE % In-Tolerance at Time of Test:	95	
MTE Deviation at Time of Cal:	**-2.81**	mV
Total Cal Process Uncertainty:	**0.789**	mV
Date of MTE Calibration:	4/19/2008	
Estimated MTE Bias from Cal Results:	**-2.0324**	mV
MTE Bias Uncertainty at Time of Cal:	**0.6710**	mV
UUT CFAR at Time of Test:	10.1308	%

As Table G-2 shows, the MTE attribute is measured at 2.81 mV below nominal. The UUT end item false accept risk is computed to be 10.1308%. If this risk is considered unacceptable, some quality control action, such as the recall of the end item, may be necessary.

G.4   Cases with Unknown X

In these cases, we need to compute $CFAR$ for a $N(0, u_x)$ population tested using an MTE from a $N(X| u_y)$ population. For this analysis, we compute $CFAR$ as the probability that the UUT was out-of-tolerance, given that it was observed to be in-tolerance during testing. From Section 4.2 of this Annex, this is just

$$CFAR = 1 - P_{xy} / P_y,$$

where

$$P_{xy} = \frac{1}{\sqrt{2\pi}} \int_{-L_1/u_x}^{L_2/u_x} \left[\Phi\left(\frac{L_1 + u_x\zeta + b}{u_y} \right) + \Phi\left(\frac{L_2 - u_x\zeta - b}{u_y} \right) - 1 \right] e^{-\zeta^2/2} d\zeta ,$$

$$P_y = \Phi\left(\frac{L_1 + b}{u}\right) + \Phi\left(\frac{L_2 - b}{u}\right) - 1,$$

and

$$u = \sqrt{u_x^2 + u_y^2}\ .$$

G.5   Estimating MTE Bias and Bias Uncertainty at $t = 0$

The bias in the MTE is estimated from calibration results using the Bayesian methodology alluded to earlier. With this methodology, we start with *a priori* information prior to measurement and use the measurement result to update and refine our knowledge. The *a priori* information consists of an estimate of the bias uncertainty of the MTE attribute prior to calibration and an estimate of the uncertainty in the calibration process.[66]

For purposes of computation we denote the bias in the MTE attribute by the variable z and the measured value obtained by calibration by the variable w. We assume *a priori* that z is normally distributed with zero mean and w is normally distributed with mean z:

$$f_z(z) = \frac{1}{\sqrt{2\pi}u_z}e^{-z^2/2u_z^2},$$

and

$$f_w(w\,|\,z) = \frac{1}{\sqrt{2\pi}u_w}e^{-(w-z)^2/2u_w^2},$$

where u_z and u_w are, respectively, the a priori estimates for the MTE bias uncertainty and the total calibration process uncertainty.

We now employ Bayesian methods to seek the conditional distribution for z, given a calibration result Z. This distribution is developed from the relation

$$h(z\,|\,Z) = \frac{f_w(w\,|\,z)f_z(z)}{f_w(Z)},$$

where

$$f_w(Z) = \int_{-\infty}^{\infty} f_w(w\,|\,Z)f_z(Z)\,dw\,.$$

Substituting the above pdfs for z and w in these expressions yields

$$f_w(w\,|\,z)f_z(z) = Ce^{-G(Z)}\exp\left[-\frac{1}{2u_z^2}\left(1+r_c^2\right)\left(z - \frac{Zr_c^2}{1+r_c^2}\right)^2\right],$$

where C is a constant, $G(Z)$ is a function of Z only, and r_c is the ratio

$$r_c = \frac{u_z}{u_w}\,.$$

---

[66] See Appendix C.

The pdf $f_w(Z)$ is obtained by integrating $f_w(w|z)f_z(z)$:

$$f_w(Z) = Ce^{-G(Z)}\sqrt{\frac{2\pi}{1+r_c^2}}u_z.$$

Dividing this expression into the expression for $f_w(w|z)f_z(z)$ gives

$$h(z|Z) = \frac{\alpha}{\sqrt{2\pi}u_z}e^{\alpha^2(z-b)^2/2u_z^2},$$

where

$$\alpha = \sqrt{1+r_c^2},$$

and

$$b = \frac{Zr_c^2}{1+r_c^2}.$$

Based on the measurement result Z, we now modify our knowledge regarding the distribution of the variable z from the pdf $h(z|Z)$. We conclude that z is normally distributed with mean b and standard deviation (standard uncertainty) u_z / α. The quantity b is the MTE bias we use in the expression for $a_\pm$. The uncertainty in this bias at the time of MTE calibration is

$$u_b(0) = u_z / \alpha.$$

G.6   Estimating MTE Bias Uncertainty at $t = T$

The uncertainty computed in Topic A is an estimate of the uncertainty in the MTE attribute value at the time of calibration. This uncertainty begins to increase from the time of measurement as a result of stresses encountered during shipping, handling, storage and general usage.

One way of looking at this is to say that, immediately following measurement, we estimate the uncertainty in the measurement to be $u_b(0)$. At some time t later, the uncertainty is $u_b(t)$. The difference between $u_b(0)$ and $u_b(t)$ is called **uncertainty growth**.

G.6.1   Fundamental Postulate

The error or bias in a subject attribute may grow with time or may remain constant. In some cases, it may even shrink. The uncertainty in this error, however, *always* grows with time since measurement. This is the **fundamental postulate** of uncertainty growth.

G.6.2   Estimating Uncertainty Growth

One way to estimate uncertainty growth is to extrapolate from in-tolerance probability vs. time data for the population to which the variable of interest belongs. Such data are referred to as *calibration history data* or *test history data*. Test or calibration history data may be fit to a reliability model from which an in-tolerance probability may be computed as a function of time.

If this is possible, then the uncertainty in the MTE attribute bias $u_b(t)$ may be computed from its $u_b(0)$ value. Specifically, if the in-tolerance probabilities at time 0 and time t are $R(0)$ and $R(t)$, respectively, and the attribute bias is normally distributed, then we can state that

$$u_b(t) = u_b(0) \frac{\Phi^{-1}[q(0)]}{\Phi^{-1}[q(t)]} \; ,$$

where

$$q(t) = \begin{cases} [1+R(t)]/2, & \text{two sided MTE specs} \\ R(t), & \text{single sided MTE specs,} \end{cases}$$

and Φ^{-1} is the inverse normal distribution function.

If reliability model coefficients are known, they can be used to compute $R(t)$. Uncertainty growth is may be estimated using the methods described in this appendix and in Appendix J. The uncertainty in b at the time of test is obtained by setting $t = T$.

G.7 Estimating Test Process Uncertainty

In this section, we determine *a priori* estimates of the UUT bias uncertainty and the uncertainty in the test process. To illustrate the method, we will assume that all biases and other errors are normally distributed with zero mean.

G.7.1 UUT Bias Uncertainty

The uncertainty in the UUT bias is obtained from the expression

$$p_{in} = \frac{1}{\sqrt{2\pi}u_x} \int_{-L_1}^{L_2} e^{-x^2/2u_x^2} dx$$

$$= \Phi\left(\frac{L_1}{u_x}\right) + \Phi\left(\frac{L_2}{u_x}\right) - 1 \; ,$$

where p_{ir} is the in-tolerance probability of the UUT as received for testing, and $\Phi(\cdot)$ is the normal distribution function.

If $L_1 \neq L_2$, then u_x is solved numerically. If $L_1 = L_2 = L$, then

$$u_x = \frac{L}{\Phi^{-1}(p)} \; ,$$

where

$$p = \begin{cases} (1 + p_{in})/2, & \text{two sided UUT specs} \\ p_{in}, & \text{single sided UUT specs.} \end{cases}$$

G.7.2 Test Process Uncertainty

G.7.2.1 Case 1 – Direct Measurement of X

Imagine that the test of the UUT attribute of interest is one involving a direct measurement of the attribute value by the MTE. Suppose also that the following errors are known to be present[67]

---

[67] The choice of applicable errors is for illustration purposes only. It does not imply that these errors are present in every measurement, nor does it imply that these errors are the only error found in practice.

b　　- MTE bias

ε_r　　- random error

ε_{op}　　- operator bias

ε_{res}　　- MTE resolution error

ε_{env}　　- error due to environmental factors

ε_{other}　　- other process errors

In this model, as in other parts of this Annex, we state the value of the measurement result as

$$Y = X + b + \varepsilon_r + \varepsilon_{op} + \varepsilon_{res} + \varepsilon_{env} + \varepsilon_{other} \, .$$

The uncertainty in Y is computed by applying the variance operator to this expression

$$
\begin{aligned}
u_y^2 &= \mathrm{var}\left(X + b + \varepsilon_r + \varepsilon_{op} + \varepsilon_{res} + \varepsilon_{env} + \varepsilon_{other} \right) \\
&= \sigma_b^2 + u_r^2 + u_{op}^2 + u_{res}^2 + u_{env}^2 + u_{other}^2 \\
&\quad + 2\rho_{b,r} u_b u_r + 2\rho_{b,op} u_b u_{op} + 2\rho_{b,res} u_b u_{res} + 2\rho_{b,env} u_b u_{env} + 2\rho_{b,other} u_b u_{other} \\
&\quad + 2\rho_{r,op} u_r u_{op} + 2\rho_{r,res} u_r u_{res} + 2\rho_{r,env} u_r u_{env} + 2\rho_{r,other} u_r u_{other} \\
&\quad + 2\rho_{op,res} u_{op} u_{res} + 2\rho_{op,env} u_{op} u_{env} + 2\rho_{op,other} u_{op} u_{other} \\
&\quad + 2\rho_{res,env} u_{res} u_{env} + 2\rho_{res,other} u_{res} u_{other} + 2\rho_{env,other} u_{env} u_{other} \, ,
\end{aligned}
$$

where ρ_{ij} is the correlation coefficient for the ith and jth test process errors. In most cases, of direct measurement, the process errors are statistically independent. This means that the correlation coefficients are zero and the above yields

$$
\begin{aligned}
u_y &= \sqrt{\mathrm{var}\left(X + b + \varepsilon_r + \varepsilon_{op} + \varepsilon_{res} + \varepsilon_{env} + \varepsilon_{other} \right)} \\
&= \sqrt{\sigma_b^2 + u_r^2 + u_{op}^2 + u_{res}^2 + u_{env}^2 + u_{other}^2} \, .
\end{aligned}
$$

G.7.2.2　Case 2 – Multivariate Measurement of X

If the test of the UUT attribute involves measuring n components $x_1, x_2, \cdots, x_n$ that comprise the variables of an equation for Y

$$Y = Y(x_1, x_2, \mathrm{L}\, , x_n) \, .$$

The error model is then written[68]

$$\varepsilon_Y = \sum_{i=1}^{n} \left(\frac{\partial Y}{\partial x_i} \right) \varepsilon_i \, ,$$

where ε_i denotes total measurement process error for the measurement of the variable x_i, $i = 1, 2, \cdots, n$. With this model, the test process uncertainty is written

---

[68] See Handbook Annex 3.

$$u_y^2 = \sum_{i=1}^{n} \left(\frac{\partial Y}{\partial x_i} \right)^2 u_i^2 + 2 \sum_{i=1}^{n-1} \sum_{j=i+1}^{n} \rho_{ij} \left(\frac{\partial Y}{\partial x_i} \right) \left(\frac{\partial Y}{\partial x_j} \right) u_i u_j \,,$$

where u_i is the total process uncertainty associated with the measurement of x_i and ρ_{ij} is the correlation coefficient for ε_i and ε_j. In applying this model, the bias uncertainty of each variable would need to be estimated as described in Section G.5.

Appendix G References

[G-1] Castrup, H., "Intercomparison of Standards: General Case," SAI Comsystems. D.O. 4M03, Dept. of the Navy Contract N00123-83-D-0015, March 1984.

[G-2] Castrup, H., "Analytical Metrology SPC Methods for ATE Implementation," *Proc. NCSL Workshop & Symposium*, Albuquerque, August 1991.

[G-3] Cousins, R., "Why Isn't Every Physicist a Bayesian," *Am. J. Phys.*, Vol 63, No. 5, May 1995.

[G-4] Jackson, D., "A Derivation of Analytical Methods to be Used in a Manometer Audit System Providing Tolerance Testing and Built-In Test," SAIC/MED-TR-830016-4M112/005-01, Dept. of the Navy, NWS Seal Beach, October 1985.

[G-5] Jackson, D., "Analytical Methods to be Used in the Computer Software for the Manometer Audit System," SAIC/MED-TR-830016-4M112/006-01, Dept. of the Navy, NWS Seal Beach, October 1985.

[G-6] Jackson, D. and Dwyer, S., "Bayesian Calibration Analysis," *Proc. NCSLI Workshop & Symposium*, Washington D.C., 2005.

[G-7] NCSL, *Establishment and Adjustment of Calibration Intervals*, Recommended Practice RP-1, National Conference of Standards Laboratories, January 1996.

Appendix H: Risk-Based End of Period Reliability Targets

H.1 Background

H.1.1 General Methodology

The approach involves establishing false accept risk confidence limits based on a computed variance in this risk. The variance in the risk follows from the variances in the projected reliabilities of an MTE attribute and a UUT attribute; each referenced to times elapsed since calibration. The variance in the projected reliability is computed from the variance-covariance matrices for the parameters of the models used to project the reliabilities.

Obviously, this is a difficult problem to solve. Accordingly, we will employ a very simple model along with a number of simplifying assumptions and conditions.

H.1.2 Application

The reliability targets developed in this appendix are applicable to equipment or system attributes. The implementation of such targets presupposes the capability to estimate corresponding calibration intervals and, in the case of multifunction devices, to establish item recall cycles based on these intervals.

H.2 Appendix H Notation

The notation used in this appendix is described in Table H-1.

Table H-1. Variables Used in Estimating Risk-Based Reliability Targets.

Variable		Description
UUT	-	a unit under test drawn randomly from an equipment population.
UUT attribute	-	a specific attribute of the UUT under consideration.
MTE	-	an instrument drawn randomly from a measuring or test equipment population used to calibrate the UUT.
MTE attribute	-	the attribute used as a reference in calibrating the UUT attribute.
$e_{UUT,b}$	-	the UUT attribute bias.
δ	-	MTE estimate (measurement) of the UUT attribute bias.
$\pm L$	-	the tolerance limits for the UUT attribute.
L	-	the range $[-L, L]$, i.e., the tolerance limits, for $e_{UUT,b}$.
$\pm l$	-	the tolerance limits for the MTE attribute.
R_x	-	the measurement reliability of the UUT attribute at the time of calibration.
R_y	-	the measurement reliability of the MTE attribute at the time of calibration.
$u_{UUT,b}$	-	the pre-test standard deviation for $e_{UUT,b}$ at the time of calibration.
u_{cal}	-	the standard deviation for δ at the time of calibration.[69]
a	-	the nominal accuracy ratio between the UUT attribute and the MTE attribute.
$f(e_{UUT,b})$	-	the probability density function (pdf) for $e_{UUT,b}$.
$f(\delta\|e_{UUT,b})$	-	the conditional pdf for δ given a specific value of $e_{UUT,b}$.

---

[69] The time of test or calibration refers to the date that the UUT attribute is tested or calibrated.

Variable		Description
P_y	-	$P(\delta \in L)$. The probability that a UUT attribute bias will be found within L.
P_{xy}	-	$P(e_{UUT,b} \in L, \delta \in L)$. The joint probability that a UUT attribute bias will lie within L and be observed to lie within L.
$UFAR$	-	unconditional false accept risk.[70]
$UFAR_{max}$	-	the maximum allowable false accept risk.
I_x	-	the UUT calibration interval.
I_y	-	the MTE calibration interval.
R^*	-	the UUT end-of-period reliability target.
MLE	-	maximum likelihood estimation. A method of estimating the parameters of a mathematical function that maximizes the likelihood of obtaining the data that has been observed \ for the function.

H.3   Assumptions

Several simplifying assumptions are made to facilitate the development of the topic. The penalization to more case-specific assumptions is straightforward, although somewhat more tedious.

H.3.1   Normal Distributions with Zero Population Bias

In this appendix, both $e_{UUT,b}$ and δ are assumed to be normally distributed with pdfs given by

$$f(e_{UUT,b}) = \frac{1}{\sqrt{2\pi}u_x} e^{-e_{UUT,b}^2/2\sigma_x^2} ,$$ (H-1)

and

$$f(\delta \mid e_{UUT,b}) = \frac{1}{\sqrt{2\pi}u_y} e^{-(\delta - e_{UUT,b})^2/2u_y^2} .$$ (H-2)

H.3.2   No Guardbands[71]

The acceptance region for $e_{UUT,b}$ is taken to be L, i.e., the tolerance limits $\pm L$.

H.3.3   False Accept Risk Definition

The definition of false accept risk employed in this appendix is the joint probability

$$UFAR = P(e_{UUT,b} \notin L, \delta \in L),$$ (H-3)

that is, the probability that the value x will be both out-of-tolerance and observed to be in-tolerance.

H.3.4   Single-Parameter Reliability Functions

It will be assumed that both the UUT attribute reliability model and MTE attribute reliability model are both non-intercept exponential models. Extension to more complex models can readily be made.

---

[70] See Chapter 3.

[71] See Chapter 5.

H.3.5 Zero Measurement Process Uncertainty

It will be assumed that the only measurement uncertainty in the UUT calibration is the bias uncertainty of the MTE attribute. Other uncertainties, such as those due to repeatability error, reproducibility error, resolution error, etc. can be included by a simple modification to the expression for u_y.

H.3.6 Use of "True" Reliabilities

In developing expressions for false accept risk and related quantities, true values for UUT and MTE attribute reliabilities will be used rather than observed values.[72] This does not detract from the method, however, since the UUT attribute reliability will be replaced later with a reliability target and the calibrating MTE in-tolerance probability will be set at its AOP value.

H.4 False Accept Risk Computation

Expanding the expression for *UFAR*, we have

$$
\begin{aligned}
FA &= P(\delta \in \mathsf{L}) - P(e_{UUY,b} \in \mathsf{L}, \delta \in \mathsf{L}) \\
&= P_y - P_{xy},
\end{aligned}
\tag{H-4}
$$

where the notation for the probability terms has been simplified for ease of development. The first term on the right-hand side is computed from

$$
\begin{aligned}
P_y &= \int_{-\infty}^{\infty} f(e_{UUT,b}) de_{UUT,b} \int_{-L}^{L} f(\delta \mid e_{UUT,b}) d\delta \\
&= 2\Phi\left(\frac{L}{u_A}\right) - 1,
\end{aligned}
\tag{H-5}
$$

where $\Phi(\cdot)$ is the cumulative normal distribution function, and

$$
u_A = \sqrt{u_x^2 + u_y^2} \ .
\tag{H-6}
$$

The second term on the right-hand side of Eq. (H-4) is computed from

$$
\begin{aligned}
P_{xy} &= \int_{-L}^{L} de_{UUT,b} f(e_{UUT,b}) \int_{-L}^{L} dy f(y \mid e_{UUT,b}) \\
&= \frac{1}{2\pi\sigma_x\sigma_y} \int_{-L}^{L} e^{-e_{UUT,b}^2/2u_x^2} de_{UUT,b} \int_{-L}^{L} e^{-(\delta - e_{UUT,b})^2/2u_y^2} d\delta \\
&= \frac{1}{2\pi} \int_{-L/u_x}^{L/u_x} d\xi\, e^{-\xi^2/2} \int_{-(L/u_y + u_x/u_y\xi)}^{L/u_y - u_x/u_y\xi} d\zeta\, e^{-\zeta^2/2} \ .
\end{aligned}
\tag{H-7}
$$

H.5 Reliability Dependence

Using Eq. (H-1), we get

---

[72] See Appendix E.

- 176 -

$$R_x = 2\Phi\left(\frac{L}{u_x}\right) - 1. \tag{H-8}$$

for the reliability of $e_{UUT,b}$ at the time of calibration. This yields a ratio φ_x that will be useful throughout the present development

$$\varphi_x \equiv \Phi^{-1}\left(\frac{1 + R_x}{2}\right). \tag{H-9}$$

Likewise, from Eq. (H-2) at the referenced time, we have

$$R_y = 2\Phi\left(\frac{l}{u_y}\right) - 1, \tag{H-10}$$

from whence we get another useful ratio φ_y

$$\varphi_y \equiv \Phi^{-1}\left(\frac{1 + R_y}{2}\right). \tag{H-11}$$

Defining the nominal accuracy ratio between the UUT attribute and the MTE attribute as

$$a = L / l, \tag{H-12}$$

we can write Eq. (H-7) as

$$\varphi_y = \frac{l}{u_y} = \frac{l}{L}\frac{L}{u_y} = \frac{L}{au_y},$$

and

$$\frac{L}{u_y} = a\varphi_y. \tag{H-13}$$

Employing Eqs. (H-9) and (H-13) in Eqs. (H-5) and (H-6), yields

$$\frac{L}{u_A} = \frac{a\varphi_y}{\sqrt{1 + (a\varphi_y / \varphi_x)^2}}. \tag{H-14}$$

Since the ratio L / u_A can be expressed in terms of functions of the reliabilities R_x and R_y, then, as Eq. (H-5) shows, so can P_y:

$$P_y = 2\Phi\left(\frac{a\varphi_y}{\sqrt{1 + (a\varphi_y / \varphi_x)^2}}\right) - 1. \tag{H-15}$$

We now turn to Eq. (H-7) and rewrite the integration limits in terms of the functions φ_x and φ_y. These limits are combinations of the terms L / u_y and $(u_x / u_y)\xi$.

We already have $L / u_y = a\varphi_y$ in Eq. (H-13), and, from Eqs. (H-9) and (H-11), we get

$$\frac{u_x}{u_y} = a\varphi_y / \varphi_x. \tag{H-16}$$

Hence, Eq. (H-7) becomes

$$P_{xy} = \frac{1}{2\pi} \int\limits_{-\varphi_x}^{\varphi_x} e^{-\xi^2/2} d\xi \int\limits_{-a\varphi_y(1+\xi/\varphi_x)}^{a\varphi_y(1-\xi/\varphi_x)} e^{-\zeta^2/2} d\zeta .$$

(H-17)

Carrying out the integration over ζ allows us to express P_{xy} as a single integral:

$$P_{xy} = \frac{1}{\sqrt{2\pi}} \int\limits_{-\varphi_x}^{\varphi_x} \left\{ \Phi\left[a\varphi_y(1+\xi/\varphi_x)\right] + \Phi\left[a\varphi_y(1-\xi/\varphi_x)\right] \right\} e^{-\xi^2/2} d\xi \; - 1 .$$

(H-18)

Combining Eqs. (H-15) and (H-18) in Eq. (H-4) then yields

$$UFAR = 2\Phi\left(\frac{a\varphi_y}{\sqrt{1+(a\varphi_y/\varphi_x)^2}} \right)$$

$$- \frac{1}{\sqrt{2\pi}} \int\limits_{-\varphi_x}^{\varphi_x} \left\{ \Phi\left[a\varphi_y(1+\xi/\varphi_x)\right] + \Phi\left[a\varphi_y(1-\xi/\varphi_x)\right] \right\} e^{-\xi^2/2} d\xi .$$

(H-19)

H.6   Variance in the False Accept Risk

We obtain the variance in $UFAR$ in Eq. (H-19) using small error theory:[73]

$$\text{var}(UFAR) = \left(\frac{\partial UFAR}{\partial \varphi_x} \right)^2 \text{var}(\varphi_x) + \left(\frac{\partial UFAR}{\partial \varphi_y} \right)^2 \text{var}(\varphi_y) .$$

(H-20)

Note that there is no covariance term in this expression. This is because we will later be expressing φ_x and φ_y as functions of two different reliability functions whose MLE parameters are arrived at independently.

We now obtain the variance terms in Eq. (H-20) also through the use of small error theory:

$$\text{var}(\varphi_x) = \left(\frac{d\varphi_x}{dR_x} \right)^2 \text{var}(R_x) ,$$

(H-21)

and

$$\text{var}(\varphi_y) = \left(\frac{d\varphi_y}{dR_y} \right)^2 \text{var}(R_y) .$$

(H-22)

As for the derivatives in Eqs. (H-21) and (H-22), it is easy to show that

$$\frac{d\varphi_x}{dR_x} = \sqrt{\frac{\pi}{2}} e^{\varphi_x^2/2} \quad \text{and} \quad \frac{d\varphi_y}{dR_y} = \sqrt{\frac{\pi}{2}} e^{\varphi_y^2/2} .$$

(H-23)

---

[73] Small error theory expresses the error in a multivariate quantity as a Taylor series expansion to first order in the error terms. The square root of the variance in this error is the measurement uncertainty in the value of the quantity [H-3]. See also Handbook Annex 3.

So, Eqs. (H-21) and (H-22) become

$$\text{var}(\varphi_x) = \frac{\pi}{2} e^{\varphi_x^2} \text{var}(R_x) \text{ and } \text{var}(\varphi_y) = \frac{\pi}{2} e^{\varphi_y^2} \text{var}(R_y). \tag{H-24}$$

Before moving on to developing expressions for $\text{var}(R_x)$ and $\text{var}(R_y)$, we pause to obtain the derivatives in Eq. (H-20). We first define functions $w_\pm$ as

$$w_\pm = a\varphi_y(1 \pm \xi / \varphi_x). \tag{H-25}$$

We then have, after a little algebra and the application of Leibnitz's rule,

$$\frac{\partial}{\partial \varphi_x} UFAR = \frac{a\varphi_y / \varphi_x}{\pi} \int_0^{\varphi_x} \left(e^{-w_+^2/2} - e^{-w_-^2/2} \right) \xi e^{-\xi^2/2} d\xi - \sqrt{\frac{2}{\pi}} \left[\Phi(w_+) + \Phi(w_-) \right] e^{-\varphi_x^2/2}, \tag{H-26}$$

and

$$\frac{\partial}{\partial \varphi_y} UFAR = -\frac{1}{\pi\varphi_y} \int_0^{\varphi_x} \left(w_+ e^{-w_+^2/2} + w_- e^{-w_-^2/2} \right) e^{-\xi^2/2} d\xi. \tag{H-27}$$

H.7   Variance in the Reliability Functions

The variances in R_x and R_y are approximated using the variance-covariance matrices obtained by MLE fits of reliability functions to R_x and R_y [H-1]. To illustrate, we will assume that, at the time of calibration of the UUT, we can write these functions as non-intercept exponential models:[74]

$$\hat{R}_x = e^{-\lambda_x t}, \tag{H-28}$$

and[75]

$$\hat{R}_y = e^{-\lambda_y \tau}, \tag{H-29}$$

where t is the time elapsed since the UUT attribute was last calibrated and τ is the time elapsed since the MTE attribute was last calibrated. With these simple models, we have

$$\begin{aligned} \text{var}(R_x) &\cong \text{var}(\hat{R}_x) \\ &= t^2 e^{-2\lambda_x t} \text{var}(\lambda_x) \\ &= t^2 \hat{R}_x^2 \text{var}(\lambda_x), \end{aligned} \tag{H-30}$$

and

$$\begin{aligned} \text{var}(R_y) &\cong \text{var}(\hat{R}_y) \\ &= \tau^2 e^{-2\lambda_y \tau} \text{var}(\lambda_y). \\ &= \tau^2 \hat{R}_y^2 \text{var}(\lambda_y). \end{aligned} \tag{H-31}$$

---

[74] Other useful reliability models are available. See Reference [H-1], Appendix A and Appendix E.

[75] In applying the method, we use these functions to compute approximate values for φ_x and φ_y.

H.8   False Accept Risk Confidence Limit

Combining Eqs. (H-30) and (H-31) with (H-21) - (H-24) and substituting in Eq. (H-20) gives

$$\text{var}(UFAR) \cong \frac{\pi}{2}\left(\frac{\partial}{\partial \varphi_x} UFAR\right)^2 e^{\varphi_x^2} t^2 \hat{R}_x^2 \, \text{var}(\lambda_x) + \frac{\pi}{2}\left(\frac{\partial}{\partial \varphi_y} UFAR\right)^2 e^{\varphi_y^2} \tau^2 \hat{R}_y^2 \, \text{var}(\lambda_y), \quad \text{(H-32)}$$

where the derivatives are given in Eqs. (H-26) and (H-27). Note that this expression is of the form

$$\text{var}(UFAR) = c_x \, \text{var}(\lambda_x) + c_y \, \text{var}(\lambda_y),$$

which we encounter frequently in dealing with multivariate measurement uncertainty analysis problems.[76] Thus, since we have independence between λ_x and λ_y, we can use the Welch-Satterthwaite relation to obtain the degrees of freedom for the variance in $UFAR$ in terms of the degrees of freedom for the variances in λ_x and λ_y. Let these be ν_x and ν_y, respectively. Then, we have

$$\nu \cong \frac{\text{var}^2(UFAR)}{\dfrac{c_x^2 \, \text{var}^2(\lambda_x)}{\nu_x} + \dfrac{c_y^2 \, \text{var}^2(\lambda_y)}{\nu_y}}. \quad \text{(H-33)}$$

This result, together with the variance in $UFAR$, can be used to obtain an upper confidence limit for $UFAR$. If the confidence level for this upper limit is $1 - \alpha$, then we have

$$UFAR_{upper} = UFAR_0 + t_{\alpha,\nu}\sqrt{\text{var}(UFAR)},^{[77]} \quad \text{(H-34)}$$

where $t_{\alpha,\nu}$ is a single-sided t-statistic and $UFAR_0$ is a "nominal" level of risk, computed from Eq. (H-19).

H.8.1   Developing the Confidence Limit

The confidence limit for $UFAR$ is obtained by setting $UFAR_{upper}$ equal to the maximum allowable false accept risk in Eq. (H-34) and solving the equation

$$UFAR + t_{\alpha,\nu}\sqrt{\text{var}(UFAR)} - UFAR_{max} = 0, \quad \text{(H-35)}$$

where $UFAR$ is given in Eq. (H-19) and var($UFAR$) in Eq. (H-32). The variables involved in the solution are R_x, R_y, t, τ, a, var(λ_x) and var(λ_y). Note that knowledge of λ_x and λ_y is not required. This is because we will employ a solution strategy the takes these parameters out of the picture.

In this strategy, we assume that the UUT is calibrated at the end of its interval. Thus we set R_x equal to the reliability target R^* and set t equal to the UUT calibration interval I_x

$$\begin{aligned} R_x &= R^* \\ t &= I_x. \end{aligned} \quad \text{(H-36)}$$

---

[76] See Appendix A.

[77] The validity of this expression will be discussed later.

We can't make the same assumptions for R_y and τ since τ could lie anywhere within the MTE calibration interval. Instead, we use the average of R_y over time and set τ equal to the time within the interval that R_y is equal to its average value. Since we're assuming that R_y follows a non-intercept exponential model, its average is equal to the square root of its end-of-period value and the corresponding time τ is half the MTE interval.[78] Assuming that the MTE has the same reliability target as the UUT, we put these considerations together and write

$$R_y = \sqrt{R^*}$$
$$\tau = I_y / 2. \tag{H-37}$$

Substituting R_x and R_y for the projections of R_x and R_y in Eq. (H-32) and using Eqs. (H-36) and (H-37) we get

$$\text{var}(UFAR) \cong \frac{\pi}{2}\left(\frac{\partial}{\partial \varphi_x} UFAR\right)^2 e^{\varphi_x^2} I_x^2 \left(R^*\right)^2 \text{var}(\lambda_x)$$
$$+ \frac{\pi}{8}\left(\frac{\partial}{\partial \varphi_y} UFAR\right)^2 e^{\varphi_y^2} I_y^2 R^* \text{var}(\lambda_y) \tag{H-38}$$

This is the expression to use in Eq. (H-35). In arriving at the solution, we redefine φ_x and φ_y

$$\varphi_x = \Phi^{-1}\left(\frac{1+R^*}{2}\right),$$

$$\varphi_y = \Phi^{-1}\left(\frac{1+\sqrt{R^*}}{2}\right),$$

and express $\hat{I}_x$ and $\hat{I}_y$ in terms of R^*

$$\hat{I}_x = -\frac{1}{\lambda_x}\ln R^*, \text{ and } \hat{I}_y = -\frac{1}{\lambda_y}\ln R^*.$$

H.8.2   Implementing the Solution

The solution of Eq. (H-35) for a reliability target R^* is arrived at iteratively. The process stops when the absolute value of the left-hand side of the equation is less than or equal to some preset small value ε. Suppose this occurs at the nth iteration, i.e., we obtain some value R^*_n.

Once we have this value, we use it to set calibration intervals for the UUT and MTE attributes. For the example in this appendix, we have

$$I_x = -\frac{1}{\lambda_x}\ln R^*_n, \text{ and } I_y = -\frac{1}{\lambda_y}\ln R^*_n.$$

H.9   A Note of Caution

The weak link in the risk-based approach described in this appendix is Eq. (H-35). This equation assumes that $UFAR$ is normally distributed. From Eq. (H-19), we see that this is not the case.

---

[78] These adjustments are valid only for reliabilities governed by the non-intercept exponential model.

Moreover, in a generalization of the risk-based approach, obtaining a distribution for *UFAR* in closed form is not feasible.

However, given φ_x and φ_y and a distribution for ξ, a distribution for *UFAR* could be assembled using Monte Carlo methods. Once this distribution is constructed, it could be used without recourse to a t-statistic to obtain an upper limit for *UFAR*. Such a construction would slow the iterative process considerably, but with current and anticipated PC processing speeds, this does not present a serious obstacle.

H.10   An Alternative Method

An approach has been suggested in which the interval I_x is solved for using an expression something like

$$R^* = \hat{R}_x - t_{\alpha,\nu}\sqrt{\mathrm{var}(\hat{R}_x)}\,, \tag{H-39}$$

and

$$\hat{I}_x = -\frac{1}{\lambda_x}\ln\hat{R}_x\,. \tag{H-40}$$

where α is the same as before, and ν is the degrees of freedom for $\mathrm{var}(\hat{R}_x)$.

From Eq. (H-30), we know that

$$\mathrm{var}(\hat{R}_x) \cong t^2\hat{R}_x^2\,\mathrm{var}(\lambda_x)\,,$$

which we replace with

$$\mathrm{var}(\hat{R}_x) \cong \hat{I}_x^2\hat{R}_x^2\,\mathrm{var}(\lambda_x)\,. \tag{H-41}$$

Substituting in Eq. (H-39) gives

$$R^* = \hat{R}_x - t_{\alpha,\nu}\hat{I}_x\hat{R}_x\sqrt{\mathrm{var}(\lambda_x)}\,. \tag{H-42}$$

We iteratively search for a value for $\hat{R}_x$ that satisfies Eq. (H-42). Once this solution is found, we compute the calibration interval I_x using

$$I_x = -\frac{1}{\lambda_x}\ln R_n\,,$$

where R_n is the solution for the *n*th iteration for $\hat{R}_x$.

While this approach is simpler than the risk-based approach proposed in this appendix, it is not satisfying from the standpoint of managing to a known level of risk, although the attained level of risk could be computed after the fact using Eq. (H-19).

Consequently, the risk-based approach is recommended. It should be mentioned, however, that the extension of the approach to accommodate more complicated reliability models, non-normal distributions and non-zero measurement process uncertainties, while conceptually straightforward, is not trivial in terms of labor.

Presumably, the task would be made somewhat easier through the application of matrices and the numerical computation of derivatives. The advantage of this is that, once the algorithms are written, the job is over. From then on it's just a matter of feeding and crunching.

H.10.1 Developing a Binomial Confidence Limit for R_x

As an alternative to Eq. (H-39), we develop an upper binomial confidence limit R_u for R^* computed from z "pseudo successes" out of m "pseudo trials." Then R_u is obtained by solving[79]

$$\sum_{k=0}^{z} \binom{m}{k} R_u^k (1 - R_u)^{m-k} = \begin{cases} \alpha, & 0 < z < m \\ 1, & z = m. \end{cases}$$

We obtain m and z by taking advantage of the properties of the binomial distribution. If z is binomially distributed, the variance in an "observed" probability $\hat{p} = z / m$ is given by

$$\mathrm{var}(\hat{p}) = \frac{p(1-p)}{m},$$

where p is the underlying probability for successful outcomes. Using the form of this expression, we establish the number of pseudo trials as

$$m = \frac{R^*(1 - R^*)}{\mathrm{var}(\hat{R}_x)},$$

and the number of pseudo successes as

$$z = mR^*.$$

Using Eq. (H-41) we have

$$m \cong \frac{1 - R^*}{\hat{I}_x^2 R^* \mathrm{var}(\lambda_x)},$$

where $\hat{I}_x$ is given by Eq. (H-40). Once m, z and R_u are computed, we obtain the calibration interval I_x from

$$I_x = -\frac{1}{\lambda_x} \ln R_u.$$

Note that, in using R_u to set I_x, we say we are setting an interval that corresponds to producing observed reliabilities of R^* or higher with approximately $1 - \alpha$ confidence.

Appendix H References

[H-1] NCSL, *Establishment and Adjustment of Calibration Intervals*, Recommended Practice RP-1, National Conference of Standards Laboratories, January 1996.

[H-2] Abramowitz, M, and Stegun, I, *Handbook of Mathematical Functions*, U.S. Dept. of Commerce Applied Mathematics Series **55**, 1972. Also available from www.dlmf.nist.gov.

---

[79] Actually, we would use the incomplete beta function defined as [H-2]

$$I_p(x, n - x + 1) = \sum_{k=x}^{n} \binom{n}{k} p^k (1 - p)^{n-k}.$$

[H-3] NCSLI, *Determining and Reporting Measurement Uncertainties*, Recommended Practice RP-12, NCSL International, Under Revision.

Appendix I: Set Theory Notation for Risk Analysis

I.1   Basic Notation

For risk analysis, the "sets" we refer to are just ranges of attribute values or biases and their complements. So, if an attribute bias $e_{UUT,b}$ is given the tolerance limits $-L_1$ and L_2, we say that it is in tolerance if

$$-L_1 \le e_{UUT,b} \le L_2 .$$

With set theory notation or, similarly, the notation of mathematical logic, we would define a set, in this case all attribute values between $-L_1$ and L_2, inclusive, with a designator like L. This can be defined as $L = \{ e_{UUT,b} \mid -L_1 \le e_{UUT,b} \le L_2 \}$. In this expression, the $\{\}$ brackets signify "the set of" and the $\mid$ symbol means "such that." So, this reads "L is the set of all values of $e_{UUT,b}$ such that $e_{UUT,b}$ is contained within the limits $-L_1$ and L_2."[80]

Actually, you don't really need to become familiar with this notation. It's just a convenient way to define things. Once you get a good idea what L is, if you like, you can use the statement

$$e_{UUT,b} \in L$$

in place of the statement.

$$-L_1 \le e_{UUT,b} \le L_2 .$$

All that is new in this notation is the $\in$ symbol, which means "is a part of" or "belongs to" or "is contained within." As for the complement of L, there are two simple ways to say that $e_{UUT,b}$ is not contained in L. The first is

$$e_{UUT,b} \notin L,$$

and the second is

$$e_{UUT,b} \in \overline{L} .$$

The symbol $\notin$ means "not contained in" and the bar over the set L indicates all values other than those contained in L. Other conventions can be used to signify complements, as is discussed later.

I.2   Additional Notation

There are other symbols that may be used in risk analysis discussions. For example,

$$A \subset B.$$

reads "A is a subset of B." This expression would apply, for example, if[81]

$$B = \{ e_{UUT,b} \mid L_1 \le e_{UUT,b} \le L_2 \}$$

and

---

[80] You could also define the set L using the "for all" symbol $\forall$. For example, L could be defined by the expression

$$L_1 \le e_{UUT,b} \le L_2 \; \forall \; e_{UUT,b} \in L.$$

[81] See the discussion in the first note, alluded to above.

$$A = \{ e_{UUT,b} \mid 0 \leq e_{UUT,b} \leq L_2 \}.$$

Incidentally, the expression $A \subseteq B$ means the same thing as $A \subset B$.

Other symbols of possible use are the "union" symbol $\cup$ and the "intersection" symbol $\cap$. The union of two sets A and B, written $A \cup B$, is the set of all numbers that belong to either A or B or both. The intersection of two sets A and B, written $A \cap B$, is the set of all numbers that belong to both A and B. In the probability functions we use in risk analysis, we could write the probability of A or B occurring as $P(A \cup B)$ and we could write the probability both A and B occur as $P(A \cap B)$. Incidentally, a relation that is sometimes useful is

$$P(A \cup B) = P(A) + P(B) - P(A \cap B).$$

Normally, we don't really use these set theory symbols in our discussions. Instead, we use a more expeditious notation wherein $P(A \cap B)$ is written $P(A,B)$, and the above becomes

$$P(A \cup B) = P(A) + P(B) - P(A,B).$$

Of course, if A and B are mutually exclusive, they can't both happen together and $P(A,B) = 0$.

The complement of a set can be written

$$A', \overline{A}, A^c$$

or any number of other ways an author is comfortable with. So, if we use the middle notation, we would write the probability of A occurring and B not occurring as $P(A, \overline{B})$. For example, if A is the event where a UUT attribute is in-tolerance and B represents the event that it is also observed to be in-tolerance then we have false accept risk given as

$$FRR = P(A, \overline{B}).$$

We can use the basic notation of the previous section to describe A and B in the argument of P. Let $e_{UUT,b}$ denote the value of the UUT attribute bias and let δ denote a measurement of this value. Then we can write

$$FRR = P(e_{UUT,b} \in \mathsf{L}, \delta \in \overline{\mathsf{L}}),$$

where $\mathsf{L} = \{ e_{UUT,b} \mid L_1 \leq e_{UUT,b} \leq L_2 \}$, as before. Alternatively, we could write

$$FRR = P(e_{UUT,b} \in \mathsf{L}, \delta \notin \mathsf{L}).$$

Extension of the notation to unconditional false accept risk ($UFAR$) is left as an exercise for the reader.

We also need to be able to accommodate the probability of an event occurring given that another event has occurred. For this we use the | symbol. With this symbol, the probability of A occurring, given that B has occurred is written $P(A|B)$. So, letting $\mathcal{A}$ represent the set of observed UUT attributes that were accepted during calibration or testing, we could write the probability of finding an out-of-tolerance UUT attribute in the accepted lot as

$$CFAR = P(e_{UUT,b} \notin \mathsf{L} \mid \delta \in \mathcal{A}).$$

This, of course, is the definition of conditional false accept risk.

Appendix J: Post-Test Risk Analysis

In calibration and testing, the measurement results are given either in a report of the measured value with details regarding the measurement uncertainty, in the case of standards calibration, or in terms of the results of a conformance test, as in the calibration of test equipment or the testing of end items.

Ordinarily, these results *do not* include uncertainties for the effects of transport to and from the calibrating or testing facility, effects of environmental conditions (i.e., temperature, humidity, barometric pressure, etc.) or drifts with time. In some cases, uncertainties arising from these effects may be greater than the reported uncertainty of the test or calibration. It is important to bear in mind that a test or calibration is guaranteed valid only at the time and place it was carried out – *it may not be relevant to using the calibrated or tested attribute in the user's environment.*

The following are factors to consider in overcoming this deficiency:

· Account for the response of the tested or calibrated attribute to shipping and handling stress during transport.

· Account for the effect of environment on the attribute and evaluate any effects if the local environment differs significantly from the one in which the attribute was calibrated or tested.

· Equipment attributes are not absolutely stable with time and, therefore, must be recalibrated periodically as discussed in Chapter 7 of the Handbook. This instability can be accommodated in uncertainty and risk calculations by accounting for the effect of uncertainty growth over time.

These factors are discussed in the following sections.

J.1 Stress Response

Accounting for the impact on a tested or calibrated attribute of a shipping, handling or environmental stress can be accomplished by an analysis of the attribute's response to stress.

The impact of a given stress on the value of an attribute is determined by using an appropriate response coefficient. For instance, the response coefficient for the effect of mechanical shock on a voltage source might be expressed as "0.015 μV per g" and the effect of temperature on a 2.5 psi pressure transducer might be stated as "±0.5% of 1.0% of output (0 to 50 °C)."

J.1.1 Shipping and Handling

Instruments and standards are often transported to and from calibration or testing facilities by hand, by special transport or by common carrier. For some attributes, the uncertainty in errors due to shipping and handling may be significant relative to the attribute's accuracy requirements.

The impact of the shipping and handling stresses on the measurement uncertainty obtained by calibration or testing can be assessed by identifying each relevant stress and quantifying the associated stress response.

In the assessment of the impact of stress on attribute value, a stress standard uncertainty is computed for each relevant stress. This standard uncertainty can be computed from entered stress limits, confidence level and a degrees of freedom estimate quantifying the amount of information available in determining each stress limit. Individual standard uncertainties are combined in the same manner as uncertainties for testing and calibration errors.[82] An example of such an analysis is shown below.

Calibration of HP 34420A at 10V DC. Response to Shipping and Handling Stress

UUT Attribute:
Attribute Name: 10 Volt DC Reading
Qualifier 1:
Qualifier 2:

Analysis Results:

Event Description	Stress Limits (g)	% Confidence	Stress Uncertainty (g)	Deg. Freedom	Response Coefficient (μV/g)	Response Uncertainty (μV)	Distribution
Cal Lab to Shipping Dock	0.75	95.0	0.38		0.015	0.006	Normal
Shipping Dock to Delivery Van	1.5	90.0	0.9		0.015	0.014	Normal
Delivery to Receiving Dock	2.2	90.0	1.3		0.015	0.020	Normal
Receiving Dock to User	0.75	95.00	0.38		0.015	0.006	Normal

Analysis Summary:
Stress Response Uncertainty: 0.026 uV
Distribution: Normal
Degrees of Freedom: ∞
Analysis Category: Type B

J.1.2 Usage Environment

Even if equipment are under ideal or "nominal" conditions, the environmental stress can impact the values of tested or calibrated attributes. The impact of nominal environmental stresses can be accounted for through the analysis of uncertainty growth with time. The impact of additional or unusual levels of environmental stress can be accounted for in the same manner as stresses due to shipping and handling. The following adds the responses to such environmental stresses to the report shown above.

---

[82] See Annex 3 and Appendix A of Annex 4.

Calibration of HP 34420A at 10V DC. Response to Shipping and Handling Stress

UUT Attribute:
Attribute Name: 10 Volt DC Reading
Qualifier 1:
Qualifier 2:

Analysis Results:

Event Description	Stress Limits (g)	% Confidence	Stress Uncertainty (g)	Deg. Freedom	Response Coefficient (μV/g)	Response Uncertainty (μV)	Distribution
Cal Lab to Shipping Dock	0.75	95.0	0.38		0.015	0.006	Normal
Shipping Dock to Delivery Van	1.5	90.0	0.9		0.015	0.014	Normal
Delivery to Receiving Dock	2.2	90.0	1.3		0.015	0.020	Normal
Receiving Dock to User	0.75	95.00	0.38		0.015	0.006	Normal
Ambient Temperature	25.0	95.0	11.8	16	0.005	0.059	Student's t

Analysis Summary:
Stress Response Uncertainty: 0.064 uV
Distribution: Student's t
Degrees of Freedom: 23
Analysis Category: Type A, B

J.2   Uncertainty Growth

An established tenet of analytical metrology is the *principal of uncertainty growth* which states that the uncertainty u_{cal} in the observed bias δ of a tested or calibrated UUT attribute grows with time following testing or calibration.[83] We indicate this time-dependence by writing the bias as $e_b(t)$ and the bias uncertainty as $u_b(t)$, where t indicates time elapsed since test or calibration ($t = 0$). Then, drawing from Appendix A, we see that $e_b(0) = \delta$ and $u_b(0) = u_{cal}$.[84]

There are two alternatives for computing $u_b(t)$; one that employs attributes data and one that employs variables data.

---

[83] In-depth discussions on uncertainty growth can be found in Annex 3 and Annex 5.

[84] The value of δ relevant to the discussion of uncertainty growth is the value following test or calibration. This may be a corrected value or an uncorrected value. If corrected, the appropriate value for $e_b(t)$ would be an adjusted δ and $u_b(0)$ would represent a value of u_{cal}, including any contributions to u_{cal} arising from the act of correction. Whether corrected or uncorrected, the values at time $t = 0$ will be represented by δ and u_{cal} in the following sections.

J.2.1 Attributes Data Analysis

Attributes data consist of in- or out-of-tolerance conditions for a population of UUT attribute values recorded during testing or calibration.[85] The following assumes that reliability modeling is used to analyze such data. The results of analysis include the selection of an appropriate measurement reliability (in-tolerance probability) model and solutions for the model's parameters, as described in Chapter 7 of the Handbook and in Annex 5.

The uncertainty $u_b(t)$ is computed using the value of the initial measurement uncertainty, $u_b(0)$, and the reliability model for the UUT attribute population. The measurement reliability of the UUT attribute at time t is related to the attribute's uncertainty according to

$$R(t) = \int_{-L_1}^{L_2} f_U[e_b(t)]de_b ,$$
(J-1)

where $f_U[\varepsilon_b(t)]$ is the probability density function for the attribute's bias at time t, and $-L_1$ and L_2 are the attribute's tolerance limits. For discussion purposes, we assume for the moment that $\varepsilon_b(t)$ is normally distributed with the pdf given by

$$f_U[e_b(t)] = \frac{1}{\sqrt{2\pi}u_b(t)} e^{-[e_b^2 - \mu_b(t)]/2u_b^2(t)} .$$
(J-2)

where the variable $\mu_b(t)$ represents the attributes expected bias at time t. The relationship between L_1, L_2 and μ is shown in Figure J-5. Also shown is the distribution of the population of biases for the UUT attribute of interest.

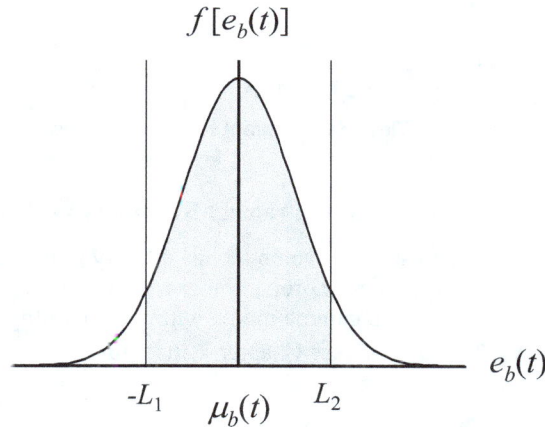

Figure J-1. Probability Density Function for UUT Attribute Bias

The shaded area represents the in-tolerance probability at time t.

We state that at a given time t, the UUT attribute's expected deviation from nominal is given by the relation

$$\mu_b(t) = \mu_0 + b(t) .$$
(J-3)

---

[85] For many testing or calibration organizations, only data recorded at the UUT item level are available. When this is the case, a UUT is considered in-tolerance only if all its tested or calibrated attributes are observed to be in-tolerance, i.e., the item is called out-of-tolerance if any single attribute is observed to be out-of-tolerance.

At the time of measurement, $t = 0$, we have $\mu_0 = \delta$ and $b(0) = 0$. The remainder of this discussion provides a method for calculating $u_b(t)$, given $u_b(0)$.

Uncertainty Growth Modeling

If we had at our disposal the reliability model for the individual measured attribute, given its initial uncertainty, we could obtain the uncertainty $u_b(t)$ by iteration or by other means. However, we usually only have information that relates to the characteristics of the reliability model for the population to which the UUT attribute belongs. This reliability model predicts the in-tolerance probability for the UUT attribute population as a function of time elapsed since measurement. It can be thought of as a function that quantifies the *stability* of the population with respect to the ability of the population members to remain in-tolerance. In this view, we begin with a population in-tolerance probability at time $t = 0$ (immediately following measurement) and extrapolate to the in-tolerance probability at time $t > 0$.

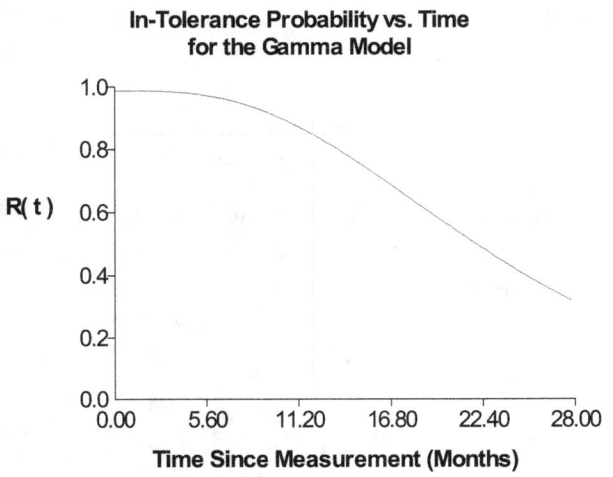

In-Tolerance Probability vs. Time for the Gamma Model

Figure J-2. Measurement Reliability vs. Time

Reliability vs. time information used to develop the uncertainty growth characteristics of the MTE attribute. The BOP and EOP in-tolerance probabilities are set at 99% and 85%, respectively. The reliability model shown is a modified gamma model with the reliability function
$R(t) = e^{-\lambda t}[1 + \lambda t + (\lambda t)^2 / 2 + (\lambda t)^3 / 6]$ (see Chapter 7 and Annex 5).

If we have recourse to an application that performs measurement reliability modeling using a comprehensive set of models,[86] we can identify the appropriate reliability model and compute its parameters. If we do not have recourse to the parameters of the reliability model, we instead utilize an elapsed time, a beginning-of-period (BOP) reliability and an end-of-period (EOP) reliability. For certain models, we must also estimate an AOP reliability. These values apply to the UUT attribute's population and are based on service history records or engineering knowledge.

---

[86] See Method S2 of Annex 5.

We next apply the reliability model obtained from these values to the individual attribute under consideration. In doing this, we operate under a set of assumptions.

1. The result of an attribute measurement is an estimate of an attribute's value or bias. This result is accompanied by an estimate of the uncertainty in the attribute's bias.

2. The uncertainty of the attribute's bias or value at time $t = 0$ (immediately following measurement) is the uncertainty of the measurement process.[87]

3. The bias or value of the measured attribute is normally distributed around the measurement result.

4. The stability of the attribute is inferred from the stability of its population. This stability is represented by the attribute population reliability model.

5. The uncertainty in the attribute's value or bias grows from its value at $t = 0$ in accordance with the reliability model of the attribute's population.[88]

Uncertainty Growth Estimation

As indicated above, uncertainty growth is estimated using the reliability model for the UUT attribute. The expressions used to compute uncertainty growth vary depending on whether the attribute tolerances are two-sided, single-sided upper or single-sided lower.

Two-Sided Cases
Using Eqs. (J-1) and (J-2) for normally distributed attribute biases with two-sided tolerance limits, the reliability function at $t = 0$ is given by

$$R(0) = \int_{-L_1}^{L_2} f_U[\varepsilon_b(0)]d\varepsilon_b$$

$$= \Phi\left(\frac{L_1 + \mu_0}{u_0}\right) + \Phi\left(\frac{L_2 - \mu_0}{u_0}\right) - 1,$$

(J-4)

where

$$u_0 \equiv u_b(0).$$

(J-5)

The parameter μ_0 is an estimate of the attribute's bias at time $t = 0$, set equal to either a sample mean or a Bayesian estimate for δ. If μ_0 is set equal to a sample mean value, u_0 is set equal to the combined uncertainty estimate for the mean value. If μ_0 is set equal to a Bayesian estimate, u_0 is set equal to the uncertainty of the Bayesian estimate.

The reliability at time $t > 0$ is given by

$$R(t) = \Phi\left[\frac{L_1 + \mu_b(t)}{u_b(t)}\right] + \Phi\left[\frac{L_2 - \mu_b(t)}{u_t(t)}\right] - 1.$$

(J-6)

---

[87] This may include an additional uncertainty due to error introduced by attribute adjustment or correction.
[88] See Eq. (J-1).

We use relations of the form of Eqs. (J-4) and (J-6) to estimate uncertainty growth. Since this growth consists of an increase in the initial uncertainty estimate, based on knowledge of the stability of the attribute population, it should not be influenced by the quantity $\mu_b(t)$. Accordingly, we construct two population reliability functions R_0 and R_t, defined by

$$R_0 = \Phi\left(\frac{L_1}{u_0}\right) + \Phi\left(\frac{L_2}{u_0}\right) - 1, \qquad \text{(J-7)}$$

and

$$R_t = \Phi\left(\frac{L_1}{u_t}\right) + \Phi\left(\frac{L_2}{u_t}\right) - 1. \qquad \text{(J-8)}$$

Next, we solve for u_0 and u_t iteratively using the algorithms in Appendix F of Annex 4. Having obtained the solutions, we write

$$u_b(t) = u_b(0)\frac{u_t}{u_0}. \qquad \text{(J-9)}$$

Once we obtain $u_b(t)$, we are on the way to solving for the in-tolerance probability at time t by using Eq. (J-6).

At this point, we need a "best" estimate for $\mu_b(t)$. If the function $b(t)$ is not known, we use the last known value of μ_b, namely μ_0, the value obtained by measurement during testing or calibration. Substituting μ_0 for μ in Eq. (J-6) we have

$$R(t) \cong \Phi\left[\frac{L_1 + \mu_0}{u_b(t)}\right] + \Phi\left[\frac{L_2 - \mu_0}{u_b(t)}\right] - 1. \qquad \text{(J-10)}$$

<u>Single-Sided Cases</u>
In cases where tolerances are single-sided, $u_b(t)$ can be determined without iteration. In these cases, either L_1 or L_2 is infinite, and Eqs. (J-7) and (J-9) become

$$R_0 = \Phi\left(\frac{L}{u_0}\right)$$

and

$$R_t = \Phi\left(\frac{L}{u_t}\right),$$

where L is equal to L_1 for single-sided lower cases and is equal to L_2 for single-sided upper cases. Solving for $u_b(t)$ yields

$$u_b(t) = u_b(0)\frac{\Phi^{-1}(R_0)}{\Phi^{-1}(R_t)}. \qquad \text{(J-11)}$$

J.2.2 Variables Data Analysis

In variables data analysis, changes in attribute values are modeled with time-dependent functions and uncertainty growth is estimated explicitly. The analysis employs as-left and as-found attribute values obtained during successive calibrations.

A methodology has been developed that employs regression analysis to model the time dependence of $\mu_b(t)$, as expressed in Eq. (J-3), and of estimating $u_b(t)$.[89] In this methodology, attribute biases are assumed to be normally distributed with variance σ^2 and mean $\mu_b(t)$, where t is the time elapsed since calibration. Regression analysis with polynomial models of arbitrary degree is applied to estimate $\mu_b(t)$. Based on the results of regression analysis, calibration intervals are determined that satisfy either EOP in-tolerance criteria or maximum allowable uncertainty criteria.

J.2.2.1 The Variables Data Model

The basic model for $\mu_b(t)$ is a generalization of Eq. (J-3)

$$\mu_b(t) = \mu_0 + \Delta(t), \tag{J-12}$$

where t is the time elapsed since calibration, $\mu_0 = y(0)$, and $\Delta(t)$ is the deviation in attribute value from μ_0 as a function of t. We model the function $\Delta(t)$ with a polynomial function $\hat{\Delta}(t)$, yielding a predicted value

$$\hat{\mu}_b(t) = \mu_0 + \hat{\Delta}(t), \tag{J-13}$$

where $\hat{\Delta}(t)$ is s-independent of μ_0, i.e., the deviation over time is not influenced by the attribute value starting point μ_0.

J.2.2.2 Uncertainty in the Projected Value

The uncertainty in a projected value of $\mu_b(t)$ is estimated as the square root of the variance of $\hat{\mu}_b(t)$

$$u_b(t) \cong \sqrt{\operatorname{var}[\hat{\mu}_b(t)]}. \tag{J-14}$$

Given the s-independence of $\Delta(t)$ and μ_0 in Eq. (J-12), we write

$$\operatorname{var}[\hat{\mu}_b(t)] = u_0^2 + s_{\Delta|t}^2, \tag{J-15}$$

where the quantity u_0 is defined as before. The quantity $s_{\Delta|t}$ is called the *standard error of the forecast*. This quantity is determined using the results of regression analysis, as will be shown later.

J.2.2.3 Regression Analysis

In modeling $\hat{\mu}_b(t)$, we employ unweighted regression analysis with observed values of Δ. The minimum data elements needed for unweighted fits are (1) the calibration service date, (2) the as-found value at the service date, (3) the previous service date and (4) the as-left value at the previous service date. Table J-1 provides an example of the kind of data that we would need to assemble from variables data service history.

---

[89] The methodology is given in Annex 3 and Annex 5.

Shown in Table J-1 are the minimum fields required for modeling attribute changes over time and uncertainty growth.

Table J-1. Example Service History Data

Service Date	As-Found Value	As-Left Value
March 29, 2003	5.173	5.073
July 11, 2003	5.324	5.048
October 5, 2003	5.158	4.993
February 17, 2004	5.292	5.126
April 27, 2004	5.226	5.024
October 17, 2004	5.639	5.208
April 2, 2005	5.611	5.451

The sampled values of $\Delta(t)$ are the differences between the as-found values and previous as-left values. The intervals between successive calibrations are called *resubmission times*. Table J-2 shows the Table J-1 data formatted for regression analysis and sorted by resubmission time.

Table J-2. Conditioned Variables Data Sample.

Resubmission Time t	$\Delta(t)$
70	0.1
86	0.11
104	0.251
135	0.299
167	0.403
173	0.615

NOTE: Data are compiled from Table J-1 and sorted by resubmission time. Imagine that we observe n pairs of as-found and as-left values with corresponding resubmission times. Let Y_i represent the as-found UUT attribute bias estimate at time X_i whose prior as-left value is y_i recorded at time x_i. We form the variables

$$\Delta_i = Y_i - y_i \qquad \text{(J-16)}$$

and

$$t_i = X_i - x_i , \qquad \text{(J-17)}$$

and write the bias drift model as

$$\hat{\Delta}_i = b_1 t_i + b_2 t_i^2 + \text{L} + b_m t_i^m , \qquad \text{(J-18)}$$

taking into account the reasonable assertion that $\Delta(0) = 0$.

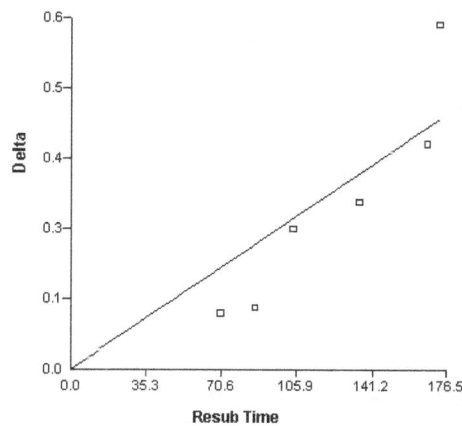

Figure J-3. First Degree Regression Fit to the Data of Table J-2

We use regression analysis on the sample of observed values of Δ and resubmission times t to solve for the coefficients b_j, $j = 1, 2, \cdots, m$. The degree of the polynomial m is a variable to be determined. Figures J-3 and J-4 show first degree ($m = 1$) and second degree ($m = 2$) polynomial regression fits to the data of Table J-2.

J.2.2.4 Projecting Attribute Values and Uncertainties

The value $\Delta(t)$ is estimated using Eq. (J-18). The uncertainty in this value is given by

$$s_{\Delta|t} = s\sqrt{1 + \text{var}[\hat{\Delta}(t)]} , \qquad (J\text{-}19)$$

where

$$s = \sqrt{\frac{RSS}{k-m}} ,$$

and RSS is the residual sum of squares, defined for a sample of k observed values by

$$\text{RSS} = \sum_{i=1}^{k} \left(\Delta_i - \hat{\Delta}_i\right)^2 . \qquad (J\text{-}20)$$

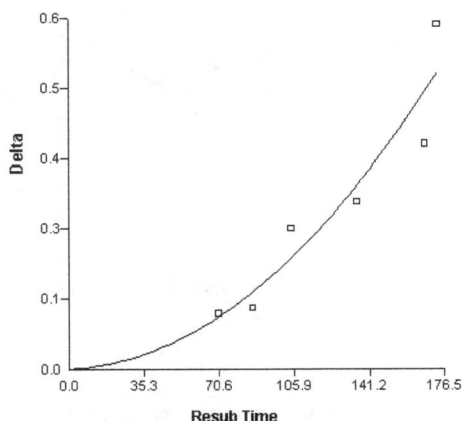

Figure J-4. Second Degree Regression Fit to the Data of Table J-2

In this expression, Δ_i is the ith observed deviation and $\hat{\Delta}_i$ is computed using Eq. (J-18). The quantity $s_{\Delta|t}$ in Eq. (J-15) is the standard error of the forecast in Eq. (J-15). It represents the uncertainty in a projected individual value of Δ, given a value of t.

The value $\mu_b(t)$ is estimated using Eq. (J-13)

$$\hat{\mu}_b(t) = \mu_0 + \hat{\Delta}(t), \tag{J-21}$$

with the uncertainty computed using Eq. (J-15)

$$u_b(t) = \sqrt{\mathrm{var}[\hat{\mu}_b(t)]}$$
$$= \sqrt{u_0^2 + s_{\Delta|t}^2} . \tag{J-22}$$

As a matter of potential future interest, the degrees of freedom ν for the estimate $u_b(t)$ is computed using the Welch-Satterthwaite relation

$$\nu = \frac{u^4(t)}{\dfrac{u_0^4}{\nu_{cal}} + \dfrac{s_{\Delta|t}^4}{n-m}} .$$

J.3   Implementation

After accounting for the uncertainty in measurement of the UUT attribute, the uncertainty due to stress response, the change in calibrated attribute value over time and the effect of uncertainty growth, it remains to estimate the risks associated with using the attribute in performing tests or calibrations on a workload item's attribute at time $t > 0$. Note that estimating the impact of stress and uncertainty growth on these risks involves the use of $\Delta(t)$ and $u(t)$, defined in Eqs. (J-13) and (J-22).

The UUT in use now becomes the MTE in the risk expressions of Chapters 3 and 4. Since the UUT attribute is now functioning as the reference attribute, it will be referred to in the following substitutions as the "MTE attribute."

The UUT bias $e_{UUT,b}$ and $u_{UUT,b}$

These are now the bias and *a priori* bias uncertainty for an attribute to be tested or calibrated by the MTE attribute. The bias and the bias uncertainty for this attribute are given in Eqs. (J-21) and (J-22), respectively.

The Calibration Uncertainty u_{cal}

The calibration uncertainty u_{cal} is obtained for all scenarios by replacing $u_{MTE,b}$ with $u_b(t)$ of Eq. (J-14) in all expressions for u_{cal}.

Appendix K: Derivation of the Degrees of Freedom Equation

An uncertainty estimate computed as the standard deviation of a random sample of measurements or determined by analysis of variance is called a *Type A estimate*. An uncertainty estimate determined heuristically, in the absence of sampled data, is called a *Type B estimate*.

The current mindset is that a Type A estimate is a "statistical" quantity, whereas a Type B estimate is not. The main reason for this is that we can qualify a Type A estimate by the amount of information that went into calculating it, whereas it is commonly believed that we can't do the same for a Type B estimate.

The amount of information used to estimate the uncertainty in a given error is called the *degrees of freedom*. The degrees of freedom is required, among other things, to employ an uncertainty estimate in computing confidence limits commensurate with some desired confidence level.

K.1   Type A Degrees of Freedom
K.1.1   Random Error

From the discussion on direct measurements, recall that the uncertainty due to random error is given by

$$u_{random} \; ; \; s_x$$
$$= \sqrt{\frac{1}{n-1}\sum_{i=1}^{n}(x_i - \overline{x})^2} \, ,$$

where x_i is the *ith* measured value in the sample, n is the sample size, $\overline{x}$ is the sample mean and s_x is the sample standard deviation. The "approximately equal" sign in this expression indicates that the measurement sample is finite. The amount of information that went into estimating the uncertainty due to random error is the degrees of freedom v_{random}. For the above estimate, it is equal to $n-1$:

$$v_{random} = n-1 \, .$$

K.2   Type B Degrees of Freedom

For a Type A estimate, the degrees of freedom is obtained as a property of a measurement sample. Since a Type B estimate is, by definition, obtained without recourse to a sample of data, we obviously don't have a sample size or other property to work with. However, we *can* develop something analogous to a sample size by applying the method described in this section.[90]

This method involves extracting what is known about a given measurement error and then converting this information into an effective degrees of freedom.[91]

---

[90] Taken from "Note on the Degrees of Freedom Equation." See www.isgmax.com.

[91] The method assumes, as in the development of most statistical tools, that measurement errors are approximately normally distributed.

Note that, to be consistent with articles written of the subject of Type B degrees of freedom, we use the notation $\sigma^2(\varepsilon)$ in this section to represent the variance in a quantity ε, i.e.,

$$\sigma^2(\varepsilon) = \text{var}(\varepsilon) = \left\langle \varepsilon^2 \right\rangle .$$

K.2.1   Methodology

The key to estimating the degrees of freedom for a Type B uncertainty estimate lies in considering the distribution for a sample standard deviation for a sample with sample size n. We know that the degrees of freedom for the standard deviation estimate is $v = n - 1$.

Let s_v represent the standard deviation, taken on a sample of size $n = v + 1$ of a $N(0, u^2)$ variable x. Given this, the quantity $v s_v^2 / u^2$ is χ^2-distributed with v degrees of freedom.

The χ^2-distribution has the pdf

$$f(x) = \frac{x^{(v-1)/2} e^{-x/2}}{2^{v/2} \Gamma\left(\dfrac{v}{2}\right)} .$$

Accordingly, we set $x = v s_v^2 / u^2$, or $s_v^2 = \left(u^2 / v\right)x$, and compute the variance in s_v^2.

$$\sigma^2\left(s_v^2\right) = \text{var}\left(s_v^2\right) = \frac{u^4}{v^2}\,\text{var}\left(x\right). \tag{K-5}$$

For a χ^2-distributed variable x, we have

$$\text{var}\left(x\right) = 2v ,$$

so that

$$\sigma^2\left(s_v^2\right) = \frac{2u^4}{v} , \tag{K-6}$$

and

$$v = 2\frac{u^4}{\sigma^2\left(s_v^2\right)} . \tag{K-7}$$

We now replace the sample variance s_v^2 with the population variance u^2 and write

$$v \; ; \; 2\frac{u^4}{\sigma^2\left(u^2\right)} . \tag{K-8}$$

To obtain the variance $\sigma^2(u^2)$, we work with the expression for the uncertainty in a normally distributed error

$$u = \frac{L}{\varphi(p)} , \tag{K-9}$$

where $\pm L$ are error containment limits, p is the containment probability and

$$\varphi(p) = \Phi^{-1}\left(\frac{1+p}{2}\right) . \tag{K-10}$$

From Eq. (K-9), we have

$$u^2 = \frac{L^2}{\varphi^2(p)} \qquad \text{(K-11)}$$

and the error in u^2 is

$$\varepsilon(u^2) \cong \left(\frac{\partial u^2}{\partial L}\right)\varepsilon(L) + \left(\frac{\partial u^2}{\partial p}\right)\varepsilon(p). \qquad \text{(K-12)}$$

Note that the variance in u^2 is synonymous with the variance in $\varepsilon(u^2)$. Hence

$$\sigma^2(u^2) = \text{var}(u^2) = \text{var}[\varepsilon(u^2)]$$

$$= \left(\frac{\partial u^2}{\partial L}\right)^2 \left\langle \varepsilon^2(L)\right\rangle + \left(\frac{\partial u^2}{\partial p}\right)^2 \left\langle \varepsilon^2(p)\right\rangle \qquad \text{(K-13)}$$

$$= \left(\frac{\partial u^2}{\partial L}\right)^2 u_L^2 + \left(\frac{\partial u^2}{\partial p}\right)^2 u_p^2,$$

where $\varepsilon(L)$ and $\varepsilon(p)$ are assumed to be s-independent, and where u_L is the uncertainty in the containment limit L and u_p is the uncertainty in the containment probability p. In this expression, the equalities

$$u_L^2 = u_{\varepsilon_L}^2 = \left\langle \varepsilon^2(L)\right\rangle$$

and

$$u_p^2 = u_{\varepsilon_p}^2 = \left\langle \varepsilon^2(p)\right\rangle$$

were used.

It now remains to determine the partial derivates. From Eq. (K-9) we get

$$\left(\frac{\partial u^2}{\partial L}\right) = \frac{2L}{\varphi^2(p)} \qquad \text{(K-14)}$$

and

$$\left(\frac{\partial u^2}{\partial p}\right) = -\frac{2L^2}{\varphi^3(p)}\frac{d\varphi}{dp}. \qquad \text{(K-15)}$$

The derivative $d\varphi/dp$ is obtained easily. We first establish that

$$\frac{1+p}{2} = \Phi[\varphi(p)]$$

$$= \frac{1}{\sqrt{2\pi}}\int_{-\infty}^{\varphi(p)} e^{-\zeta^2/2}.$$

Taking the derivative of both sides of this expression yields

$$\frac{1}{2} = \frac{1}{\sqrt{2\pi}}e^{-\varphi^2(p)/2}\frac{d\varphi}{dp},$$

from which we get

- 202 -

$$\frac{d\varphi}{dp} = \sqrt{\frac{\pi}{2}}\, e^{\varphi^2(p)/2}. \qquad\qquad (K\text{-}16)$$

Substituting in Eq. (K-15) gives

$$\left(\frac{\partial u^2}{\partial p}\right) = -\frac{2L^2}{\varphi^3(p)}\sqrt{\frac{\pi}{2}}\, e^{\varphi^2(p)/2}. \qquad\qquad (K\text{-}17)$$

Combining Eqs. (K-17) and (K-14) in Eq. (K-13), yields

$$\sigma^2(u^2) = \frac{4L^4}{\varphi^4(p)}\left[\frac{u_L^2}{L^2} + \frac{1}{\varphi^2(p)}\frac{\pi}{2}e^{\varphi^2(p)}u_p^2\right]. \qquad\qquad (K\text{-}18)$$

Substituting Eq. (K-18) in Eq. (K-8) and using Eq. (K-9) yields

$$\nu \; ; \; \frac{1}{2}\left[\frac{u_L^2}{L^2} + \frac{1}{\varphi^2(p)}\frac{\pi}{2}e^{\varphi^2(p)}u_p^2\right]^{-1}. \qquad\qquad (K\text{-}19)$$

K.2.2 Comparison with Eq. G3 of the GUM

Appendix G of the ISO Guide to the Expression of Uncertainty in Measurement (the GUM) [K-1] provides an expression for the degrees of freedom for a Type B estimate

$$\nu \; ; \; \frac{1}{2}\frac{u^2}{\sigma^2(u)}. \qquad\qquad (K\text{-}20)$$

From Eq. (K-9), we have

$$\varepsilon(u) \; ; \; \left(\frac{\partial u}{\partial L}\right)\varepsilon(L) + \left(\frac{\partial u}{\partial p}\right)\varepsilon(p)$$

$$= \frac{1}{\varphi(p)}\varepsilon(L) - \frac{L}{\varphi^2(p)}\frac{d\varphi}{dp}\varepsilon(p). \qquad\qquad (K\text{-}21)$$

Substituting from Eq. (K-16) gives

$$\varepsilon(u) \; ; \; \frac{1}{\varphi(p)}\varepsilon(L) - \frac{L}{\varphi^2(p)}\sqrt{\frac{\pi}{2}}\, e^{\varphi^2(p)/2}\varepsilon(p).$$

Applying the variance operator, we have

$$\sigma^2(u) = \text{var}(u) = \text{var}[\varepsilon(u)] = \left\langle \varepsilon^2(u) \right\rangle$$

$$= \left(\frac{\partial u}{\partial L} \right)^2 \left\langle \varepsilon^2(L) \right\rangle + \left(\frac{\partial u}{\partial p} \right)^2 \left\langle \varepsilon^2(p) \right\rangle$$

$$= \left(\frac{\partial u}{\partial L} \right)^2 u_L^2 + \left(\frac{\partial u}{\partial p} \right)^2 u_p^2 \qquad \text{(K-22)}$$

$$= \frac{1}{\varphi^2(p)} u_L^2 + \frac{L^2}{\varphi^4(p)} \frac{\pi}{2} e^{\varphi^2(p)} u_p^2$$

$$= \frac{1}{\varphi^2(p)} \left[u_L^2 + \frac{L^2}{\varphi^2(p)} \frac{\pi}{2} e^{\varphi^2(p)} u_p^2 \right].$$

Substituting from Eqs. (K-9) and (K-22) in Eq. (K-20) gives

$$\nu \; ; \; \frac{1}{2} \left[\frac{u_L^2}{L^2} + \frac{1}{\varphi^2(p)} \frac{\pi}{2} e^{\varphi^2(p)} u_p^2 \right]^{-1}. \qquad \text{(K-23)}$$

Comparison of Eq. (K-23) with Eq. (K-19) shows that using either Eq. (K-8) derived in this note or Eq. G3 of the GUM yields the same result.

K.2.3   Estimating u_L and u_p

Three formats are in use for estimating Type B degrees of freedom.[92]

Format 1
In Format 1, the containment probability is

$$p = C / 100.$$

The uncertainty is computed from the containment limits $\pm L$ and an inverse function computed from the containment probability according to

$$u = \frac{L}{\varphi(p)} \; ,$$

where, as before,

$$\varphi(p) = \Phi^{-1}\left[(1+p)/2\right] \; .$$

If error limits ΔL and Δp can be surmised for L and p, respectively, then the degrees of freedom are computed from[93]

$$\nu = \frac{3\varphi^2 L^2}{2\varphi^2(\Delta L)^2 + \pi L^2 e^{\varphi^2}(\Delta p)^2} \; .$$

---

[92] These formats are embodied in the software applications UncertaintyAnalyzer [9], Uncertainty Sidekick [10] and the Type B Degrees of Freedom Calculator [11].

[93] This result and others in this section are derived in [K-2].

Note that if ΔL and Δp are set to zero, the Type B degrees of freedom become infinite.

Format 2

In Format 2, the containment probability is $p = n / N$, where N is the number of observations of a value and n is the number of values observed to fall within $\pm L$ ($\pm \Delta L$). For this format, we use the relation

$$\nu = \frac{3\varphi^2 L^2}{2\varphi^2 (\Delta L)^2 + 3\pi L^2 e^{\varphi^2} p(1-p)/n} \ ,$$

where the quantity $p(1 - p) / N$ is the maximum likelihood estimate of the standard deviation in n, given N observations or "trials."

Format 3

Format 3 is a variation of Format 2 in which the variable C is stated in terms of a percentage of the number of observations N. In this format, $p = C / 100$. The degrees of freedom are given as with Format 2:

$$\nu = \frac{3\varphi^2 L^2}{2\varphi^2 (\Delta L)^2 + 3\pi L^2 e^{\varphi^2} p(1-p)/n} \ .$$

K.2.4   Degrees of Freedom for Combined Estimates
K.2.4.1   Statistically Independent Errors

For discussion purposes, we reiterate Eq. (K-8)

$$\nu \; ; \; 2\frac{u^4}{\sigma^2 \left(u^2 \right)} .$$

While this expression is ordinarily applied to estimating the degrees of freedom for an estimate of the uncertainty in the value of a given quantity obtained by direct measurement, it can also be used to estimate the degrees of freedom for a combined estimate, as will be seen below.

Combined Uncertainty

Imagine that the total error in measurement is the sum of two s-independent errors whose uncertainties are u_1 and u_2. Then the variance in the total error is given by

$$u^2 = u_1^2 + u_2^2 . \tag{K-24}$$

Applying the variance operator to Eq. (K-24) gives

$$\sigma^2(u^2) = \sigma^2(u_1^2) + \sigma^2(u_2^2) . \tag{K-25}$$

Let ν_1 and ν_2 represent the degrees of freedom for the estimates u_1 and u_2, respectively. Then, by Eq. (K-8),

$$\sigma^2 \left(u_1^2 \right) \; ; \; 2\frac{u_1^4}{\nu_1} , \text{ and } \sigma^2 \left(u_2^2 \right) \; ; \; 2\frac{u_2^4}{\nu_2} . \tag{K-25}$$

Using these results in Eq. (K-25) yields

$$\sigma^2\left(u^2\right) = 2\left(\frac{u_1^4}{v_1} + \frac{u_2^4}{v_2}\right). \tag{K-27}$$

Substituting Eq. (K-27) in Eq. (K-8) gives

$$v = \frac{u^4}{\dfrac{u_1^4}{v_1} + \dfrac{u_2^4}{v_2}}, \tag{K-28}$$

Welch-Satterthwaite

Extending Eq. (K-28) to the combined uncertainty in an error comprised of a linear sum of n s-independent errors yields

$$v = \frac{u^4}{\displaystyle\sum_{i=1}^{n}\frac{u_i^4}{v_i}}, \tag{K-29}$$

which is the widely used Welch-Satterthwaite relation.

K.2.4.2 Correlated Errors

Consider now a case where the measurement error is the sum of two correlated errors whose correlation coefficient is ρ_{12}. The variance in the total error is given by

$$u^2 = u_1^2 + u_2^2 + 2\rho_{12}u_1u_2. \tag{K-30}$$

From Eq. (K-30) and the variance addition rule, we have

$$\begin{aligned}
\sigma^2(u^2) = \sigma^2(u_1^2) + \sigma^2(u_2^2) + 4\rho_{12}^2\sigma^2(u_1u_2) \\
+ 2\operatorname{cov}(u_1^2, u_2^2) + 4\rho_{12}\operatorname{cov}(u_1^2, u_1u_2) + 4\rho_{12}\operatorname{cov}(u_2^2, u_1u_2).
\end{aligned} \tag{K-31}$$

Eq. (K-31) shows that we need to examine a possible correlation between the uncertainty estimates u_1 and u_2. These uncertainties are each presumably obtained using some method or prescription and, possibly, a sample of data. In some cases, the uncertainty estimates u_1 and u_2 are s-independent. In others, s-independence may not apply. In what follows, we set $E(u_1) = E(u_2) = 0$.[94]

S-Independent Uncertainty Estimates

We now develop a Welch-Satterthwaite relation for cases where, although the errors may be correlated, the uncertainty estimates are not. To do this, we first need to express the covariance terms and the cross-product $\sigma^2(u_1u_2)$ term in Eq. (K-31) as quantities that can be computed using the relations developed in this appendix.

For s-independent u_1 and u_2, the cross-product term is given by

---

[94] Since the uncertainties follow a χ^2 distribution, this is not strictly justified. We apply this artificial rule because what we are after are variances that represent variances in the "errors" in the uncertainty estimates.

$$\sigma^2(u_1u_2) = E(u_1u_2)^2 - E^2(u_1u_2)$$
$$= E(u_1^2)E(u_2^2) - E^2(u_1)E^2(u_2)$$
$$= E(u_1^2)E(u_2^2)$$
$$= \sigma^2(u_1)\sigma^2(u_2).$$

The first covariance term in Eq. (K-31) is

$$\text{cov}(u_1^2, u_2^2) = E\left(\left[u_1^2 - E(u_1^2)\right]\left[u_2^2 - E(u_2^2)\right]\right).$$

If u_1 and u_2 are s-independent then

$$E\left(\left[u_1^2 - E(u_1^2)\right]\left[u_2^2 - E(u_2^2)\right]\right) = E\left[u_1^2 - E(u_1^2)\right]E\left[u_2^2 - E(u_2^2)\right],$$

and

$$\text{cov}(u_1^2, u_2^2) = 0$$

The second covariance terms is

$$\text{cov}(u_1^2, u_1u_2) = E\left(\left[u_1^2 - E(u_1^2)\right]\left[u_1u_2 - E(u_1u_2)\right]\right)$$
$$E\left[u_1^3u_2 - u_1^2E(u_1u_2) - E(u_1^2)u_1u_2 + E(u_1^2)E(u_1u_2)\right]$$
$$= E(u_1^3u_2) - E(u_1^2)E(u_1u_2)$$
$$= E(u_1^3)E(u_2) - E(u_1^2)E(u_1)E(u_2)$$
$$= 0.$$

For the third covariance,

$$\text{cov}(u_2^2, u_1u_2) = E(u_1)E(u_2^3) - E(u_2^2)E(u_1)E(u_2)$$
$$= 0.$$

With these results, Eq. (K-31) becomes

$$\sigma^2(u^2) = \sigma^2(u_1^2) + \sigma^2(u_2^2) + 4\rho_{12}^2\sigma^2(u_1)\sigma^2(u_2). \tag{K-32}$$

By Eqs. (K-8) and (K-26), this becomes

$$\sigma^2(u^2) = 2\frac{u_1^4}{\nu_1} + 2\frac{u_2^4}{\nu_2} + 4\rho_{12}^2\sigma^2(u_1)\sigma^2(u_2).$$

Using this expression we can write the Welch-Satterthwaite relation for the degrees of freedom of a total uncertainty estimate for a combined error with correlated component errors or error sources

$$\nu = \frac{u^4}{\displaystyle\sum_{i=1}^{n}\frac{u_i^4}{\nu_i} + 2\sum_{i=1}^{n-1}\sum_{j=i+1}^{n}\rho_{ij}^2\sigma^2(u_i)\sigma^2(u_j)}, \tag{K-33}$$

where u^4 is determined from u^2, given in Eq. (K-30), and $\sigma^2(u)$ is given in Eq. (K-22).

Appendix K References

[K-1]   ANSI/NCSL Z540-2-1997, *U.S. Guide to the Expression of Uncertainty in Measurement*, National Conference of Standards Laboratories, Boulder, 1997.

[K-2]   Castrup, H., "Estimating Category B Degrees of Freedom," Measurement Science Conference, Anaheim, January 21, 2000.